A Tactical Guide to Science Journalism

A Tactical Guide to Science Journalism

Lessons From the Front Lines

Edited by

Deborah Blum, Ashley Smart, and Tom Zeller Jr.

Production Editor

Scott Veale

OXFORD
UNIVERSITY PRESS

OXFORD
UNIVERSITY PRESS

Oxford University Press is a department of the University of Oxford. It furthers the University's objective of excellence in research, scholarship, and education by publishing worldwide. Oxford is a registered trade mark of Oxford University Press in the UK and certain other countries.

Published in the United States of America by Oxford University Press
198 Madison Avenue, New York, NY 10016, United States of America.

Library of Congress Cataloging-in-Publication Data
Names: Blum, Deborah, 1954– editor. | Smart, Ashley, editor. |
Zeller Jr.,Tom, editor.
Title: A Tactical guide to science journalism : lessons from the front lines / edited by Deborah Blum, Ashley Smart, and Tom Zeller Jr.
Description: New York, NY : Oxford University Press, [2022] |
Includes bibliographical references and index.
Identifiers: LCCN 2021057624 (print) | LCCN 2021057625 (ebook) |
ISBN 9780197551509 (paperback) | ISBN 9780197551523 (epub) |
ISBN 9780197551530 (online)
Subjects: LCSH: Sports journalism.
Classification: LCC PN4784.S6 T33 2022 (print) | LCC PN4784.S6 (ebook) |
DDC 070.4/49796—dc23/eng/20220222
LC record available at https://lccn.loc.gov/2021057624
LC ebook record available at https://lccn.loc.gov/2021057625

DOI: 10.1093/oso/9780197551509.001.0001

1 3 5 7 9 8 6 4 2

Printed by Marquis, Canada

With gratitude to the Alfred P. Sloan Foundation and Program Director Doron Weber for their support of this book and for their long-standing dedication to promoting the best in science journalism.

Contents

Contributors

Nsikan Akpan
Health & Science Editor
New York Public Radio
USA

Julia Belluz
Senior Health Correspondent
Vox
USA

Deborah Blum
Director
Knight Science Journalism Program
Massachusetts Institute of Technology
USA

Brooke Borel
Senior Editor
Undark Magazine
USA

Helen Branswell
Senior Writer, Infectious Diseases and
Global Health
STAT News
USA

Iván Carrillo
Freelance Journalist
Mexico

Alicia Chang
Deputy Editor, Health and Science
Associated Press
USA

Ian Cheney
Director and Producer
Wicked Delicate Films
USA

Jeffery DelViscio
Chief Multimedia Editor
Scientific American
USA

Nadia Drake
Contributing Writer
National Geographic
USA

Katherine Eban
Contributing Editor
Vanity Fair
USA

Martin Enserink
International News Editor
Science Magazine
Netherlands

Dan Fagin
Professor
Carter Institute of Journalism
New York University
USA

Andrada Fiscutean
Freelance Journalist
Romania

Azeen Ghorayshi
Health and Science Reporter
The New York Times
USA

Pallab Ghosh
Science Correspondent
BBC
Great Britain

James Glanz
Reporter
The New York Times
USA

Elana Gordon
Reporter
Public Radio International
USA

Sujata Gupta
Reporter, Social Sciences
Science News
USA

Matthew Hutson
Contributing Writer
The New Yorker
USA

Sarah Kaplan
Science Reporter
The Washington Post
USA

Maggie Koerth
Senior Science Reporter
FiveThirtyEight
USA

Thomas Lin
Editor in Chief
Quanta Magazine
USA

Apoorva Mandavilli
Health and Science Reporter
The New York Times
USA

Betsy Mason
Freelance Journalist
USA

Rod McCullom
Freelance Journalist
USA

Megan Molteni
Staff Science Writer
STAT News
USA

Michael Morisy
Founder, MuckRock
USA

Esther Nakkazi
Founder, Health Journalists Network
Uganda
Uganda

Liz Neporent
Director, Social Media and Community
WebMD/Medscape
USA

Rachel Nuwer
Freelance Journalist
USA

Ivan Oransky
Distinguished Writer in Residence
Arthur Carter Journalism Institute
New York University
USA

Jennifer Ouellette
Senior Reporter
Ars Technica
USA

Jason Penchoff
Senior Director, Physician Communications
Memorial Sloan Kettering Cancer Center
USA

Antonio Regalado
Senior Editor, Biomedicine
MIT Technology Review
Massachusetts Institute of Technology
USA

Sabriya Rice
Knight Chair in Health and Medical
Journalism
University of Georgia
USA

Charles Seife
Professor
Arthur L. Carter Journalism Institute
New York University
USA

Ashley Smart
Associate Director
Knight Science Journalism Program
Massachusetts Institute of Technology
USA

Richard Stone
Senior Science Editor
Howard Hughes Medical Institute
USA

Kate Travis
Freelance Journalist
USA

Bina Venkataraman
Editor at Large
The Boston Globe
USA

Dan Vergano
Science Reporter
BuzzFeed News
USA

Paige Williams
Staff Writer
The New Yorker
USA

Tom Zeller Jr.
Editor in Chief
Undark Magazine
USA

Kim Zetter
Freelance Journalist
USA

Introduction

The veteran journalist Tim Radford, who headed up the science desk at the United Kingdom's *Guardian* newspaper for more than two decades, was once interviewed by a government committee charged with investigating the fragile relationship between "science and society." In a lengthy report submitted to the House of Lords in February 2000, the committee noted that the public's faith in both science and government had been shaken over the preceding years—in part by an outbreak of bovine spongiform encephalopathy, colloquially known as "mad cow disease." This and the swift rise of biotechnology, the burgeoning internet age, and other fast-moving manifestations of human ingenuity, it was determined, were creating an air of anxiety and mistrust.

Public perception of science was a key issue, the authors stated, as was the disregard many scientists had for the social implications of their research. Trust in the government's ability to distill science into sound policy was also waning, the report found. And uncertainty and risk—two concepts that are fundamental to the scientific enterprise—were deemed to be poorly understood by the general public.

Blame for this state of affairs was well distributed, but the final chapter of the report zeroed in on the sometimes-fractious relationship between scientists and the media. Radford attempted to illuminate those tensions through the tale of an ill-fated press conference he'd attended—one designed to reveal new data on Earth's core. The researcher heading up the briefing, it seemed, had offhandedly mentioned to the scrum of reporters that one of his experiments involved a 6-inch naval gun mounted in his laboratory. The journalists were subsequently full of questions and curiosity about the weapon—How was it used? Why?—and the scientist would never be able to steer things back to his findings. Radford called it an exquisite clash of cultures.

"He actually wanted to tell us about the science of the center of the Earth," he told the committee, but "we were after a story, which is something entirely different."

If you are holding this book in your hands (or reading it electronically), you are no doubt in search of stories, too—ones with a special emphasis on science. You are also likely to have noted, maybe with a bit of disquiet, how

Deborah Blum, Ashley Smart, and Tom Zeller Jr., *Introduction* In: *A Tactical Guide to Science Journalism*. Edited by: Deborah Blum, Ashley Smart, and Tom Zeller Jr., Oxford University Press. © Oxford University Press 2022. DOI: 10.1093/oso/9780197551509.003.0001 little has changed in the ongoing dance between science and the press over the last two decades. In some respects, that relationship has become even more strained as the science embedded in our lives—and buffeted by our politics, our economics, and our culture—becomes ever more complex.

Today, the public is deeply divided over issues that hinge on questions of science, from genetic engineering to the rise of artificial intelligence, and everywhere in between: the efficacy of vaccines, the rise of bioterror, the value of space exploration, the looming challenges of pandemics, the presence of extraterrestrial life, the search for new energy sources, the fluidity of gender, and the increasingly dire impacts of climate change.

These are just a few of the areas that now divide communities, animate elections, and sometimes leave even the most well-meaning reporters scratching their heads.

Despite such challenges, the profession of science journalism continues to thrive. Public health crises like the Covid-19 pandemic and environmental threats such as global climate change have reminded everyone—from editors to audiences—that informed and insightful reporting on these issues is vital to our understanding of the world around us. And the internet age, no longer in its infancy, has delivered a true blossoming of outlets devoted to the coverage of science—places where young writers can break into the business, learn the craft, and build a portfolio, as either staff members or freelance contributors.

As the profession has flourished, so have university programs, books, and websites, providing smart and useful training and advice on the craft of science writing and the different career paths possible within it. (See our Resources section on page 333.)

What hasn't been available, until now, is a single volume devoted to what we at the Knight Science Journalism Program sometimes refer to as "Capital-J Journalism." If your aim is to better understand scientific questions and to communicate that understanding to readers in lucid prose, compelling documentaries, rich podcasts, or illuminating infographics—that's great. We feel that compulsion, too.

But journalism, whether covering politics or business, crime or health, is more than simple communication. It is, as Radford himself put it in the very first of his 25 commandments for journalists, an acknowledgment that your most important obligation, your most enduring fealty, is not to "the scientist you have just interviewed, nor the professor who got you through your degree, . . . [or] even your mother." Rather, Radford insisted, your sole allegiance as a Capital-J Journalist is to your reader, your viewer, or your listener. They turn to you for fair analysis, thoughtful and accurate explanation, and dispassionate—though not brainless—distillation of competing worldviews. Readers expect, and deserve, that you deliver this without fear or favor.

For journalists covering science—and particularly for those who are new to the beat—that can seem like a tall order, and this volume is designed to help you along the way. How do you find good science stories? How does the scientific publishing process work? How can you access public documents related to science? What is it like to cover physics, or chemistry, or computer engineering, or space science—and how will you know what's important on these beats and what's not? Was that 6-inch naval gun as important to readers as the scientist's novel findings about Earth's mantle? Maybe. Maybe not.

These are just some of the questions and quandaries that the *Tactical Guide to Science Journalism* is designed to help answer. We recognize, after all, that science is full of potential land mines that can lead journalists into errors of exaggeration, hyperbole, sensationalism, or downright inaccurate portrayals. And while many science journalists are former scientists themselves, or have extensive training in science, we think it's fair to say that many, perhaps even most, have no special research expertise. And even those journalists well versed in one aspect of science—medicine, say, or public health—can suddenly find themselves flummoxed when asked to cover machine learning or landscape ecology.

At the same time, scientists and the institutional personnel who speak for them—just like ambitious politicians and swashbuckling business tycoons—often have parochial, inward-facing interests of their own. That doesn't mean they ought to be your adversaries, of course. But if you're serious about science journalism, it does mean that you need to understand and consider not only *what* a researcher is telling you, but also *why*. An ability to sift through scientific direction—and misdirection, after all—is essential for journalists who specialize in science.

Here, we have tapped some of the world's best science journalists to provide tips on how to avoid mistakes, find experts, and otherwise navigate this variegated world. But we'd argue that the lessons offered here are equally valuable to journalists across the board who want to gain the ability to add smart science coverage to their beats, as well as for scientists curious about the ways of journalism, readers wondering about how stories are told, and more. And all of this alongside a suite of eminently practical chapters on the craft of science storytelling, the business of science book writing, the challenges of international science reporting, and the virtues of science journalism across borders.

The Knight Science Journalism Program along with its award-winning digital science publication *Undark Magazine*, place special emphasis on the importance of journalism in the public interest. *Undark*'s mission statement makes that commitment clear: "As journalists, we recognize that science can often be politically, economically and ethically fraught, even as it captures the imagination and showcases the astonishing scope of human endeavor. [We] will therefore aim to explore science in both light and shadow, and to bring that exploration to a broad, international audience."

That you are reading this book suggests that you may be animated by the same mission, and it is our sincerest hope that the *Tactical Guide to Science Journalism* will serve as both a reference and a beacon as you embark on your journey. The world needs committed, courageous science journalists, and if this book convinces you to join those ranks, then our efforts will have been worthwhile.

The Editors
Deborah Blum, Director
The Knight Science Journalism Program at MIT

Ashley Smart, Associate Director
The Knight Science Journalism Program at MIT

Tom Zeller Jr., Editor in Chief
Undark Magazine

PART I
FOUNDATIONS

1

How Science Works

Nsikan Akpan

Nsikan Akpan is health and science editor at New York Public Radio and runs WNYC/Gothamist's health and science desk. He was previously a science editor at National Geographic, *where he oversaw Covid-19 coverage as well as other topics in science, health, and technology. Before that, he worked at* PBS NewsHour, *where he co-created the award-winning video series* ScienceScope. *In 2021, Nsikan's alma mater, Bard College, granted him the John Dewey Award for Distinguished Public Service. He shared a 2020 News Emmy for the* PBS NewsHour *series* Stopping a Killer Pandemic *and in 2019 received a George Foster Peabody Award for the* PBS NewsHour *series* The Plastic Problem. *Nsikan has also worked for* NPR, Science News, Science Magazine, *KUSP Central Coast Public Radio, and the* Santa Cruz Sentinel, *and was a writer at the Center for Infection and Immunity at Columbia University. He holds a doctorate in pathobiology from Columbia University and is an alumnus of the science communication program at the University of California, Santa Cruz.*

As children, we're taught to objectify science. Grade school textbooks define the field in concrete terms like "the scientific method," "series of experiments," or "people who wear lab coats."

But science in practice is better described as a mentality. This lesson didn't become apparent until I began transitioning my career from biomedical research to journalism in 2013. I had spent the previous 6 years investigating how neurons die after a person suffers a stroke—learning the minutia of apoptosis, a process wherein our cells commit suicide as a defense mechanism.

Gradually, my life approach began to shift after months and months of deciphering the natural world. The scientific method calls on a person to make repeated observations—ones that are thorough and impartial. After a while, I developed a compulsion to interrogate almost everything beyond what's presented on the surface. I felt compelled to apply statistical logic to election results. (How many people were surveyed in that mayoral poll? Had they reached a representative population? What was the margin of error?) If a

Nsikan Akpan, *How Science Works* In: *A Tactical Guide to Science Journalism.* Edited by: Deborah Blum, Ashley Smart, and Tom Zeller Jr., Oxford University Press. © Oxford University Press 2022. DOI: 10.1093/oso/9780197551509.003.0002 subway app said the next train arrived in 12 minutes, I wondered how often it was wrong and how many commutes would be needed to ever feel confident about the pattern.

Journalism similarly tries to uncover how things transpire. When a news-worthy event occurs, we document as many perspectives as possible—whether from experts or lay citizens—so that audiences can comprehend the experience without actually being there.

These overlapping mentalities explain why science and journalism are natural bedfellows. Science writers go a step beyond relaying "what happened" and dig into the "how" and the "why." Science writing can explain why megafires keep striking California as dutifully as it explains lead contamination in a water supply. If journalism strives to uncover objective facts, what better field to report on than science, which is steeped in repeated observations and empirical data?

Fully explaining how science works would take longer than a doctorate, but let's look at some simple ways to infuse a scientific mentality into journalism training and news reporting.

Not All Papers Are Trustworthy

The most common form of science news centers on discoveries that appear in academic papers.

As a scientist, I see the word "paper" and think of those that appear in academic journals like *Nature* or *Science*. But the same process underlies most reports, whether they come from a policy institute or government agency. Conveying one's findings is the last stage of any mission that sets out to answer a question (hypothesis building and testing), makes observations, and assembles data.

Different publications come with varying levels of validity. Academic journals rely on peer review, by which an independent panel of experts comments and judges a draft before publication. This industry standard ostensibly adds rigor, so that journals publish higher quality information. Preprints—manuscripts that have yet to undergo peer review—represent the first stage of this process: the assemblage of information into article form.

The Covid-19 pandemic cemented a growing trend of uploading preprints to public databases for broader critiques and dissemination. Policy groups and organizations, meanwhile, depend on their long-standing versions of preprints: white papers or summary reports.

All of these mediums can be tainted by bias or become subject to corrections and retractions when errors are made or misinformation is disseminated under the guise of legitimacy. Most preprint databases lack rigorous gatekeepers and can therefore be weaponized to spread misinformation. Policy reports can obviously carry the biases of the group's larger agenda.

While it is often (justly) criticized as being insular and subject to abuse, the peer-review process represents the most valid form of achieving a scientific presentation. Much like assigning multiple editors to look over an investigative feature, peer reviewers inject objectivity into the pursuit of communicating a study's motivation, methods, results, and conclusions. When the reviews are conducted independently and in a manner that avoids conflicts of interest, the process can help filter out results that are biased or unfounded. (If you're looking for a good book on this topic, try *Rigor Mortis* by longtime NPR health reporter Richard Harris.)

How to Read a Science Paper

Academic literature can be intimidating when you're starting as a scientist or journalist. Reading over the jargon in scientific papers can feel like learning a new language, not to mention all the charts, graphs, and numbers. Each science discipline seems to come with its own rules and terminology, so it's no surprise that approaching these papers can feel daunting.

Luckily, fledgling journalists have a handy guide for unpacking studies: news articles. One of the first assignments in journalism school involves learning about the inverted pyramid. It's the standard way to write a news story.

This news-writing format calls for a punchy, barebones synopsis at the top of the story. That's typically followed by a little background on what led up to the event. You will next find a "nutgraf"—a brief description of everything else to come in the story, including takeaways. After that, the writer typically unpacks the ins-and-outs of what took place, charting a stepwise descent to supplemental facts. The final lines might also offer a conclusion, such as future expectations (which are immaterial to the news itself) or an overarching point of view that puts the news event into a broader context.

Let's now pivot to a scientific study. You start with a straightforward summary called an abstract. The Introduction section sketches out past studies that influenced this study before ending with what the new research has found. The Methods section walks readers through the ins and outs of how the research was conducted. The Results section offers up the deliverance, the endpoint of the journey. Meanwhile, the Discussion section provides the context for where the study sits in the field and the ever-important limitations of the research. It may also sketch out the broader implications of the findings.

Once you become familiar with these two blueprints, transposing the essential information from a study into the gist of a standard news report becomes easier. And this guide works for not only a print piece but also TV or audio broadcasts.

A Journal Club for Every Newsroom

It sounds basic, but reading and discussion are the keys to learning how science works.

The best science, health, and environmental journalists are often the ones who can sit down, read a set of papers, and tell you a story about what they found—without distorting the studies. When I encounter scientific misinterpretations in newsrooms, they're most often due to a reporter or editor not taking the time to read background material.

As a result, after mentoring newsroom interns, speaking to aspiring science journalists, and editing numerous stories by science and general reporters, I believe our profession would benefit from holding journal clubs as part of training programs and editorial activities.

Journal clubs were a common thread throughout my research experiences in academia, whether they involved a lab section of an undergraduate class or a weekly session for graduate students. A journal club is a regular meeting where people agree to read the same scientific article and then discuss it. Much like a book club, you set an agenda on what to deliberate, and its success thrives on being inquisitive. No question is too silly for a journal club.

My first journal club felt laborious. I remember googling nearly every scientific term in the papers for my first few meetings. Yet over time, my mind started to discern the common language that researchers use across different fields.

My most memorable conversations from journal clubs have always centered on a paper's statistical content. The clubs forced me to digest the particulars behind each bar chart or graph. Figure legends call for brevity, and through the repetition of a journal club, you quickly learn to identify the shorthand being used by scientists. Taking an intro-level statistics class can teach you about p values, hazard ratios, and analysis of variance, but journal clubs allow you to witness those scientific tools in action.

Partake in enough journal clubs, and the upsides and flaws of papers will start to reveal themselves automatically. You'll soon be able to quickly recognize an experiment with too few participants to make firm statistical conclusions, for example.

You don't need a classroom setting to hold a journal club. Freelancers can meet to discuss major papers that come out. The journal clubs don't even need to be labeled as journal clubs. I'll often assign group projects where a pair of reporters or a team will cooperate to cover a major study. Here's the first part of this process: "Hey, here's a paper. Go read it. Let's meet in 2 days to discuss what it's about."

The Allure of Scientists

When I worked as a science producer for the *PBS NewsHour* from 2015 to 2019, I co-led a media research project looking at what viewers want from their science news content. The results were broken down between early-career adults (18–35 years) and older audiences.

But we found that if you want any viewer, regardless of age, to connect with your stories and share your content—whether written, audio, or video—then you need to address three elements in the first 10 to 15 seconds of your broadcast segment or the beginning of a print story.

The first is engaging with the readers' or viewers' perception of relevance and their morality: Why should they care about this story over the millions of pieces of content coursing through Twitter, Apple News, Instagram, Google News, TikTok, YouTube, and all the other media platforms on the internet. The second is tapping into their science identity—connecting with their inner science nerd while avoiding descriptions that might make them feel dumb or ostracized. The third key element is the opening's aesthetic appeal: catching the media consumer with something enticing they don't expect to see or hear.

While the latter applies mostly to broadcast news, relevance and science identity can be captured in any story. But to do so, a journalist must look beyond the results listed in a paper and access the people behind all those experiments.

The first question I ask every scientist during my interviews is, "What was your motivation for doing this study?" It's a good conversation starter because academic research is a laborious profession that requires a bountiful amount of self-drive. Rather than make millions on the stock market or use their technocratic skills to streamline governance, scientists obsessively examine one sliver of an infinite universe. They must have a good reason for doing so, and people tend to relate to passion.

This initial question steers the interview toward discussing why the research team made certain choices rather than focusing solely on what is contained in its results. Why did you opt to study glaciers in Greenland rather than Antarctica? Why did your study on sexually transmitted infections look at teenage populations instead of senior citizens?

For example, I once interviewed a mechanical engineer about a study he did on why shoelaces come untied. His day job consisted of examining continuum mechanics and nonlinear dynamics—essentially the math behind why materials move in unpredictable ways. His work typically looked at how machines and robots operate, so I asked why the random pivot to shoelaces. The question led to the following passage in my story:

> Mechanical engineer Oliver O'Reilly began looking into this telltale problem 3 years ago, after trying to teach his young daughter to tie her shoes.
>
> "I went online and found all these helpful videos about how to tie your shoelaces," O'Reilly said. "They were wonderful and very helpful, but I also noticed there were no videos online about why your shoelaces become untied."

Picking apart the "why" (the underlying motivations for decisions made during the research process) added to the story's richness. It also allows you to pinpoint the public value of basic research that has no apparent application and describe its moral relevance.

The Scientist Family

To engage the science enthusiasts in my audience, I've found it useful to interview a broad spectrum of individuals on a research team. These groups tend to be run by principal investigators, or PIs, who are usually professors, associate professors, or assistant professors. A postdoctoral researcher or postdoc, as the title suggests, has completed their PhD but is still in a training mode. Next, labs can feature an array of students, from doctoral candidates to postbaccalaureates to undergraduates. Staff scientists are the young professionals of academia, in that they're not typically involved in pedagogy. They can range from senior staff who've completed postdocs to the undergrads who are hired as lab technicians. They're in the lab to conduct research.

While principal investigators tend to be the senior authors—and, by extension, figureheads—for a study, try to request interviews with junior staff for news features. Early-career researchers can better convey the vibrance that goes into research as well as the insecurity. Being on the front line of an endeavor to discover something new about how the world works is audacious, and with bravery comes occasional defeats. The stakes are higher for younger scientists, and those points of view can help remove the "ivory tower" stigma that sometimes tarnishes science coverage.

Finally, follow the money. Cash rules everything in academia: from which labs get to hire the most researchers to how much effort can go into an investigation. Money is also an excellent story device, given it is universally understood. Too often in science reporting, we forget to mention funders or potential conflicts of interest, both of which can have an enormous influence on research directives.

The Covid-19 pandemic offers a good example of the influence of money. Due to the global emergency, many labs—both inside and outside of biomedicine—pivoted their research to studying the coronavirus. It was the only way for some scientists to receive permission to enter their labs during the lockdowns, continue their research, and meet the project goals mandated by their grants. Others were drawn to the topic by funding agencies that put out grants to study Covid-19. These words aren't meant to downplay the altruism that likely also drove these researchers, but it is an example of how resources dictate science. Similar incentives drive tech innovations and environmental protections.

The Takeaway

Early on in my scientific journey, I had a desire to comprehend how the brain works after a close family member developed a debilitating mental illness. I was 12 years old, scared, and confused by how their behavior had suddenly pivoted.

Around this time, my middle school biology teacher introduced us to genetics. I can vividly remember her explanation of how genes can influence the structure of our organs or how we act. Genetics and neuroscience would serve as the through line for my scientific pursuits during high school, college, and my doctoral work. The more I learned about the intricacies of the mind, the less stigma defined my perceptions of mental health.

And, in a way, that is the beauty of the science mentality. Framing science as a mentality makes it more accessible to everyone. Rather than hinge on the memorization of species names and mathematical proofs, it becomes a filter through which to interpret a chaotic world.

To take this journey, you just need to know how its tools work.

2

Finding and Vetting Sources

Azeen Ghorayshi

Azeen Ghorayshi is a science reporter for The New York Times. *Previously, she worked as both a reporter and as the editor of the science desk at* BuzzFeed News. *At Buzzfeed, she frequently wrote about topics like sex and gender, personal genetics, reproductive health, the opioid epidemic, and sexism and sexual harassment in the sciences. She's won an AAAS Kavli Science Journalism Award and the Clark/Payne Award for young science journalists, and was a finalist in the Livingston Award's national reporting category. You can find her work in the 2017 edition of* Best American Science and Nature Writing.

One Monday in February 2016, I got an email with a cryptic subject line and a link to a lawsuit with no contact information.

Getting tips like this in my inbox was not an infrequent occurrence at this point. Months earlier, I'd reported the first in a series of stories about young women scientists being sexually harassed by prominent scientists, only to find that the internal systems set up to protect them more often maintained the status quo. Because formal investigations resulted in little in the way of recourse for the women involved, many of those I spoke to ended up leaving science, driven out by shame, frustration, or both.

But this tip was unusual. The link took me to a lawsuit, filed by a prominent virologist, Michael Katze, against the University of Washington, alleging that the university had violated his tenure by banning him from entering his own lab. Buried in the complaint was a partial explanation why: Katze was being investigated for hiring an employee on the condition that she have a sexual relationship with him.

To this day I still don't know who that source was, but they had passed on a hell of a tip. Months later, after I filed a records request to the university to get access to its investigation, got sued by Katze in order not to see those records, and flew to Seattle to speak with many of his current and former colleagues, we published a story. The university, it turned out, had received at least seven

Azeen Ghorayshi, *Finding and Vetting Sources* In: *A Tactical Guide to Science Journalism.* Edited by: Deborah Blum, Ashley Smart, and Tom Zeller Jr., Oxford University Press. © Oxford University Press 2022.
DOI: 10.1093/oso/9780197551509.003.0003

formal complaints about the scientist over the last decade, but had always chosen to look the other way.

Impossible to ignore was the fact that Katze had $30 million in active grants by the time we published. Eighteen months later, he became the first tenured professor in the university's history to get fired.

This story only came about because the anonymous source knew to contact me in the first place. But in a broader sense, it was because my editor at BuzzFeed News, Virginia Hughes, now at the *New York Times*, demanded that her reporters apply scrutiny to scientific authorities, exposing the ways in which science can be subject to the same distorting forces of power, money, and biases that reporters on other beats seek out.

In this case, I had no idea what this source's intentions were. But it pointed me toward useful information that I was able to run down and verify. Sources can be people you've worked with for years, new people you seek out to get information, or tipsters whose identities you may never know. Your job as a reporter is figuring out who and what to trust.

Knowing how to navigate source relationships is crucial when doing any type of reporting—whether it's accountability driven or diving into complex new science. It takes learning how to find sources and convince them to talk to you, whether they're in positions of power or are more vulnerable; how to shrewdly obtain documents; and how to develop reliable sources who you can turn to regularly as you carve out a beat. Doing these things well can help draw more sources to you, establishing a steady pipeline of stories.

Map Out the Terrain

Starting any new reporting project requires time to get your bearings and figure out who to talk to. When you're approaching a new scientific subject, that often means poring through studies and trying to map out where ideas are coming from—what's known, who's driving those conversations, and what's on the fringes.

Review papers, which summarize a body of research on a topic, can be a helpful starting point to find a broad swath of researchers who have contributed solid work on the subject you're researching. Check that they have a track record of publishing relevant work in reputable journals. And as with any type of source, always do a side search to see if you can find any other information that would raise any red flags.

Conferences can also be a way to get access to a lot of scientific sources at once, while simultaneously seeing the dynamics of a field play out in real time.

As a journalist, Hughes often tries to attend smaller conferences, where the informality can serve as a useful peek behind the curtain. "When there isn't a ton of media around, scientists are just talking more loosely and more provocatively about what their stuff means," Hughes says.

Mapping out the group of people you're reporting on is also essential for more investigative stories. For Katherine Eban, a science journalist who landed huge scoops about the disasters unfolding behind the scenes during the Trump administration in 2020, that work is essential to reporting on the federal government. "I spend a lot of time studying org charts," she says. "There are so many little subagencies—15 people who work in one little nodule of one agency—and you've got to figure all that out to try to figure out who might know something."

Ask Questions and Avoid Hype

Every source has bias. It's a reporter's job to figure out whether what they tell you is true and newsworthy. So, be open to talking to everyone.

In my own reporting on tense scientific fights—for example, on how to best treat intersex kids—I have sometimes relied on controversial sources to help pinpoint where the tensions in the story were. In this case, the medical establishment was facing a tremendous amount of pressure from activists, many of whom were born intersex and had been operated on as infants to assign them as "boys" or "girls," to change the standard of care to one centered on informed consent. This was spurring a nascent cultural shift among many younger doctors I spoke to. But speaking to the older guard was also essential. Many of them openly disparaged the shift in thinking in their field, pointing me toward hospitals that were resisting such changes.

Going out of your way to talk to people with different viewpoints can also help ensure that you're holding all your sources to account. This was especially true during the Covid-19 pandemic, where many scientific questions became intensely politicized.

New York Times science reporter Apoorva Mandavilli spent much of 2020 reporting on some of the thorniest scientific questions surrounding the pandemic, including the uncertainty about viral transmission in kids and school openings. She found that her most reliable sources were those who acknowledged that messiness up front: "I think a good part of it is looking for that over-certain person and avoiding them."

Doing so will also help you avoid one of the biggest pitfalls of science journalism: hype.

For Hughes, a turning point in her career came around 2012, when she was reporting for *Spectrum News*. "I really started to understand there was a lot of crazy hype around brain scanning, a lot of crazy hype around genetics and what it was going to teach us," and at the end of the day there weren't any good drugs being developed to treat autism, Hughes says. At the time, there was a growing shift among many journalists to start questioning scientific authorities more. "We can't be cheerleaders," Hughes said. "We have to be careful with how we hype stuff."

Social Media Is a Priceless Resource

In 2017, I started reporting a story about the growing number of heroin users who were using a plant called kratom to help them quit. The reason? Chemicals in the plant plug into some of the same receptors as opioids, a feature that made it a clear target for the Food and Drug Administration.

I found and joined several Facebook groups where people openly discussed their dosing regimens and recipes for using the plant. Listening in on those groups gave me access to communities that would otherwise be difficult to tap into: People who use drugs are, for good reason, often suspicious of outsiders. After a while, I asked whether anyone would like to speak to me for a story. I got dozens of responses and ended up meeting several people in person and publishing a story weeks later.

My biggest tip with using Facebook groups to develop sources is this: don't lurk too hard. Be up front about who you are and the fact that you will not use anyone's stories without their permission. That honesty, in the end, will help get you the access you're looking for.

Other social media platforms are also incredibly valuable. LinkedIn is an amazing way to find potential sources inside companies. Through the LinkedIn for Journalists program, you can directly message employees who you think might have the information you're seeking.

And healthcare reporters like Sarah Kliff frequently use a "call-out" strategy on Twitter or in posts, telling people what kind of stories they're looking to investigate and asking readers to reach out with their own experiences.

During the pandemic, Twitter has also emerged as an essential space to watch scientists discuss new findings in real time. But Mandavilli warns that the platform has also created a lot of dubious experts. "One unusual thing with Covid is that people really strayed far out of their lanes," she says. She needed to keep that in mind to make sure experts are "really qualified to give the kinds of quotes they were giving."

The Power of Persuasion

One of the most difficult things about developing new sources is the initial reach out. You may send a cordial email, but never hear back. Are they just too busy, or are they avoiding you? How hard do you continue pushing them to talk? And how else can you convince them?

The answers depend on what kind of story you're pursuing. If it's a story on developing research, you may be better off seeking out a wide range of sources to talk to—a junior author on the same paper, for example, or another scientist with a similar research focus. My general rule is that persistence pays off, but within reason: Nudge them once or twice, but know that if you don't hear back by then, you're probably out of luck.

For more investigative stories, persistence is absolutely necessary. People have a lot of reasons not to talk to you, even if they want to. Your goal is to tap into the reasons they may or may not have to share information and convince them it's worth it.

Admittedly, this dynamic can be uncomfortable. While sources often have a lot to lose, professionally or even legally, journalists only stand to gain from a good story. But understanding what motivates people can help you find the reasons why they may, too, ultimately want a story to be public.

"There is a lot of psychology in it," Eban says. "What do they care about? Maybe they care about you not getting it wrong? Maybe they really feel fucked over and want to dry someone out? Maybe they're truly just motivated by concern about public health and feel no one is listening to them? You have to meet them where they are."

Convincing sources to talk can also mean literally meeting them where they're at. While some sources prefer speaking over the phone; others may prefer to text. (For sensitive sources, I try to always get them to download and use end-to-end encrypted texting platforms like Signal or WhatsApp.) Others may only feel comfortable if you fly across the country to meet them in person.

In cases where you're trying to contact someone who does not want to talk to you, you may have to cold-call them (i.e., call them without them knowing you'll be doing so) or even show up at their homes unannounced. Doing so requires sensitivity since it's an invasion of their personal space. Be sure to have your opening lines ready about why you're there and need to talk. When door-knocking, I always make sure to bring a letter with my business card attached. This shows how dedicated I am to speaking to them. It also let's them know I'm an established reporter with a solid track record, and gives people the time and space to contact me later and a way to do so.

When to Use Anonymous Sources

In an ideal world, every source you're talking to, and everything they say, would be on the record and fair to use in a story. But for sensitive pieces where someone may have a lot to lose by speaking to you, that's often not the case, for good reason. Figuring out when to push sources to go on the record and when to grant them anonymity or keep them out of the story entirely is a difficult calculus that reporters often need to work out with their editors.

This was true for stories I reported on sexual harassment and assault, where some victims inevitably decided going public would ultimately come at too great a cost to their careers. It was also true for drug users, who feared law enforcement, and for my stories on medical care for transgender and intersex kids, where my sources included families and young children. It can also be true for a source in a relative position of power.

In these cases, I often offer people a chance to speak to me off the record at first, meaning nothing from that conversation will be used in a story. "Almost all of the interviews I do lately, I do not in any way push to talk on the record," Eban says. "I want to have as candid a discussion as possible—I'm just trying to understand what I'm reporting on."

Those early conversations are also when I listen carefully for hints of any documentation—from texts, DMs, emails, and receipts, to raw data, medical records, and legal documents—that could be used to support the source's account. Stressing that documents only help make a story more solid and can protect the source's anonymity can help convince people to share such valuable information. Once you've established this trust with a source, you can also go back to them to ask to speak on the record.

Ultimately, the decision whether or not to rely on unnamed sources varies case by case. "I try to always take it back to the reader: What does naming someone do for the power of the story and the credibility of the source?" Hughes says, referring to conversations she has as an editor with her reporters. "It obviously makes it more credible when you name someone—it just always does."

"But the question is," she adds, "is it worth it for whatever potential consequences there will be for the source?"

The Takeaway

In the end, building trust requires showing that you are sensitive and thoughtful with sources—and publish stories that they view as diligent and fair. This can help you gain a broader network of people who may come to you with future tips, as well as get you sources that you can turn to for information time and time again.

Each of the stories I have reported on sexual harassment in science, for example, has gained me a wider pool of sources, especially among the "whisper network" of scientists reckoning with this problem on the ground. But those relationships also require maintenance. Don't be afraid to pick up the phone and just call your sources to chat. Open-ended conversations can often yield story ideas where you're least expecting them.

3
Journals, Peer Review, and Preprints

Ivan Oransky

Ivan Oransky is editor-in-chief of Spectrum, *Distinguished Writer in Residence at New York University's Arthur Carter Journalism Institute, and cofounder of the blog* Retraction Watch. *Ivan was previously vice president of editorial at* Medscape, *global editorial director of* MedPage Today, *and executive editor of* Reuters Health, *and held editorial positions at* Scientific American *and* The Scientist. *A 2012 TEDMED speaker, he is the recipient of the 2015 John P. McGovern Medal for excellence in biomedical communication from the American Medical Writers Association, and in 2017 he was awarded an honorary doctorate in civil laws from the University of the South (Sewanee). Ivan has also taught at the City University of New York Graduate School of Journalism and has written for publications from* Nature *to* The New York Times. *From 2017 to 2021, he served as president of the Association of Health Care Journalists.*

One day in 2007, I was sitting in Reuters' Times Square offices editing a story for physicians about a study of how to treat collapsed lungs. The study had been peer reviewed, meaning two or three experts had said it was good enough to be published.

But when I looked at the study itself, something didn't sit right. The authors had enrolled more than 100 people to assess whether putting talc into the space between a collapsed lung and the chest wall—a standard treatment— would prevent future collapses. Despite only reporting on what happened to 56 of those people, they concluded that talc was 95 percent effective. (They also didn't include what is known as a control group of people who *didn't* receive talc, which is not as unusual as it should be in studies of humans.)

In a geeky fit of pique, I redid the math as if everyone who couldn't be found because of what the authors referred to as "geographical movement" had further collapsed lungs. That—referred to as an "intention-to-treat analysis"— is what researchers are *supposed* to do: Rather than assume everyone they couldn't track down had done just fine, rigorous statistical analysis requires

Ivan Oransky, *Journals, Peer Review, and Preprints* In: *A Tactical Guide to Science Journalism.* Edited by: Deborah Blum, Ashley Smart, and Tom Zeller Jr., Oxford University Press. © Oxford University Press 2022.
DOI: 10.1093/oso/9780197551509.003.0004

that they be entered in the "ineffective" column. After all, a treatment being ineffective is a frequent reason for loss to follow-up. When I did that, it turned out the success rate was just half of what the authors claimed.

The episode wasn't my awakening to the limitations of peer review. I'd had that years before, when I learned to read studies critically as a new member of the Association of Health Care Journalists, which offers great training in reading such studies. (Full disclosure: I'm now the president of the volunteer board of directors.) But it was a potent reminder of how for every great story that comes from a study in a peer-reviewed journal, there are many that weren't vetted as well as journals would like us to think they were.

Keeping the limitations of peer review in mind as you're scouring publications and reading studies will help you be a better reporter.

Where to Find Studies

Journals publish millions of papers each year, and even some rough whittling down to articles relevant to a particular reporter's beat would yield hundreds per week. How do you find studies to cover?

One source is EurekAlert, a service of the American Association for the Advancement of Science that distributes press releases from journals, universities, and other institutions. Many of these releases are embargoed—more on that later—and most link to a study. Make sure you read the whole study, not just the abstract. Public information officers at universities and publishers are typically more than happy to provide it if it's not available directly, and that's true of papers you find elsewhere, too.

Another is PubMed, a database of abstracts housed at the National Library of Medicine (NLM). Like EurekAlert, the links you'll find in PubMed may take you to the full text of a paper or may not. Journals listed in Medline, which feeds PubMed, must satisfy certain quality criteria, but be careful using journals in PubMed Central, a confusingly named database also housed at the NLM that does not do nearly as much vetting.

One of the good things about using PubMed is that you can go beyond journals that produce press releases read by all of your competitors. I still have alerts set up in PubMed for obscure subjects, including "Major League Baseball." (Yes, studies involving baseball players appear regularly.)

Google Scholar is another source, and while it's much more permissive than other indexes (reader beware), its alerts allow you to follow the work of particular authors in a way that seems more efficient than others.

Today, you can get a glimpse of what people are saying about papers if you're looking at those that are already published. Go to PubPeer.com (disclosure: I'm a volunteer member of their board of directors) to see if anyone has commented on a particular paper. And click through to the Altmetric page on papers published by journals that make use of that service, or PlumX for Elsevier journals through ScienceDirect to see what other journalism outlets have said about the paper and what people are saying about it on social media.

How Not to Get Scammed

In 2016, the Federal Trade Commission (FTC) took an unusual step: They sued a scientific publisher, OMICS, for deceptive practices, saying that: "Many articles are published with little to no peer review and numerous individuals represented to be editors have not agreed to be affiliated with the journals." The FTC prevailed in court, and OMICS was ordered to pay the U.S. government $50 million.

OMICS had long been known in scientific circles as a "predatory publisher." The term, coined by librarian Jeffrey Beall, who kept a list of publishers he considered predatory, is controversial for several reasons. One is that it assumes that researchers who publish in these journals are unwitting prey, which isn't necessarily the case. Another is that it's a binary term that doesn't differentiate between a journal's various flaws.

There is now a more standardized list of predatory journals, run by a company called Cabell's, but figuring out which journals should be on that list can be a bit of a game of Whac-A-Mole. I would suggest asking an established researcher whether they have heard of a journal in their field. If they haven't, that's a red flag that might prompt a question for the editors about how robust their peer review is.

At the end of the day, what a reporter should really care about is whether the peer review a journal claims to have done was rigorous—or even real. In short, stay away from journals like this, regardless of what they're called.

Is That Paper Worth Covering?

Today's media landscape—including high volume, at many outlets—means using heuristics (also known as mental shortcuts), to judge studies. With the caveat that any particular heuristic shouldn't be the be-all and end-all, here are some that I have found useful:

1. **Where was the study published?** High-profile journals, particularly those that put out press releases, command the lion's share of press coverage. Competition to appear in their pages could mean robust or even brutal systems to separate the wheat from the chaff. But their peer reviewers aren't always the right kinds of experts. Don't be afraid to look at specialty journals with slightly lower profiles. They may be less likely to attract headline-worthy papers, but their peer reviewers probably put studies through their paces better.

2. **Ask a statistician.** For close to two decades, I've been telling my students at New York University's Science, Health, and Environmental Reporting Program to keep a biostatistician in their back pocket. Many of them, along with epidemiologists, have soared in popularity during the Covid-19 pandemic, which makes them easier to find, but they field lots of calls from the media. It's worth developing a relationship with someone you can send a study for a quick sniff test.

3. **Look for the limitations**. All studies have limitations, but the ones I trust are the ones that spell them out. At the very least, including limitations in your work makes for a better story.

4. **Size matters**. There are no bright lines for how many samples, animals, or participants need to be in a study for it to be meaningful, although one hopes that researchers performed what's known as a power calculation to figure that out. For rare conditions, small studies can tell us a lot. Still, in general, the larger the study, the more useful it is.

5. **Exert control.** Showing that a group of people did better after a particular treatment or intervention is all good and well, but how do you know they wouldn't have improved without that treatment? That's where a control group of people who didn't receive the intervention comes in. If a paper doesn't include such a group—known as a control group—it's unlikely to be that useful or worth writing about.

6. **Correlation is not causation.** Is there some other plausible explanation for the findings? And a related question: Did the study test what matters directly or use what is known as a surrogate marker or proxy? In other

words, did it test whether people had heart attacks or died, or did it test cholesterol? Did the study look at an esoteric test or the outcome that really matters? A clinical study or psychological experiment may be carefully done but irrelevant to most people.

Get to Know Preprints

The Covid-19 pandemic saw an explosion in scientific publishing about the virus and related issues. Understandably, scientists wanted to move quickly and to announce their results just as quickly. One phenomenon that fueled the explosion was what are known as preprint servers. The pandemic was the first time so many people became aware of such servers, with names like bioRxiv and medRxiv, which allow researchers to post non–peer-reviewed manuscripts for public view rather than waiting months or even years for them to be reviewed and then published in journals.

Preprint servers were not new to the pandemic, however. ArXiv, a preprint server for physics, math, and computer science, was created in the early 1990s, and quickly became the lingua franca for researchers in those fields. Biology and medicine were much slower to adopt the approach, in part because of fears that unvetted claims about outbreaks and other serious illnesses could lead to harm.

It was not until 2017 that a journalism student of mine at New York University pitched a story about a preprint for an assignment. The study, claiming to show that a potent psychedelic called ayahuasca could treat severe depression, seemed interesting. But the fact that it hadn't yet been peer reviewed gave me pause. Sure, peer review has flaws, but at least you know that *someone* had eyeballed the work.

The student, Emiliano Rodriguez Mega, and I came to the conclusion that it would be fine to report on the preprint as long as he could vet it more thoroughly than he might for a peer-reviewed paper. That perhaps meant more sources than average or, put another way, commissioning your own peer review. Carl Zimmer did something similar for his 2010 piece in *Slate* about a paper claiming that bacteria could live on arsenic. The paper had been published in *Science*, one of the world's leading peer-reviewed journals, but the experts Zimmer spoke to were all highly critical of the findings.

Rodriguez Mega talked to lots of sources and asked the lead author of the paper what turned out to be a revealing question: Had he submitted the preprint to a journal before it was posted, and if so, what happened? It turns out that the manuscript had been rejected by a dozen journals. That context, which appeared in his finished article, told the reader an important part of the story.

During the pandemic, journalists scooped up—pun intended—preprints as soon as they were posted and ran stories on them. But few of them had done the sort of due diligence Rodriguez Mega or Zimmer did—to the disservice of their readers.

Still, it is difficult to see the preprint genie returning to the bottle, and many have turned to developing best practices for journalists who cover preprints—a healthier approach than the "don't report on anything until we tell you to" that widespread acceptance of embargoes has ingrained for so long among reporters and journals. With freedom comes responsibility, however. More than ever journalists covering science must hone their ability to vet and report on scientific findings, including finding and interviewing sources.

Understand How Embargoes Can Limit You

Today, it is the rare science reporter who has not agreed to wholesale embargoes on journal studies, and it is therefore no surprise that so much coverage of science looks like pack journalism. One rationale for embargoes is that in exchange for receiving copies of studies early, journalists will have more time to digest the findings and report. That's true if reporters use that time instead of just waiting until a few hours before the embargo lifts anyway.

But it's a Faustian bargain: Journals get to decide when journalists report on a given set of findings, and even worse, they can scare away many scientists from talking to reporters before their work is published in their pages lest their papers be rejected. It's publish or perish at work again, this time limiting reporters' ability to write about science in progress, as opposed to as it's published.

Still, given that the realities of the news industry typically require some level of volume, not to mention scoops, covering studies is one way to feed the beast. If done with forethought, it can even provide a nutritious meal. One way around embargoes is to cover preprints—but keep their limitations in mind, too.

Peer-Reviewed Papers May Be Flawed

Peer review can, when done well, add value. But lots of editors—including some who have invited me to review papers well outside of my limited expertise based on the fact that I show up in a database—are on autopilot. That is even more alarming given that some researchers have come up with elaborate ways to fool journal editors into having their friends review their papers or review them themselves. Other academics have created "citation rings" in which they boost their metrics: You cite me, I'll cite you, and our work will look as if it's having more of an impact than it really is. It's all part of "publish or perish."

The fact is that lots of flawed papers make it through peer review. About four in 10,000 papers are eventually retracted, which seems like a low number until you realize that journals, authors, and others seem to go out of their way to find reasons not to retract, leaving poisonous fruit on the tree for others to cite—which they do. It took 12 years for *The Lancet*, one of the world's top medical journals, to retract a peer-reviewed study used to falsely claim that autism was linked to vaccines.

The number of papers that turn out to be too flawed for journalists to pay attention to is much higher than four in 10,000. Well over 100 retractions of papers about Covid-19 were retracted in the first 12 months of the pandemic—a number that was about what you'd expect after a few years.

So beware. Ask that statistician to review the "breakthrough" paper someone just sent you to cover. And check out the Retraction Watch Database to see if any of the authors have had papers retracted.

The Takeaway

Scientific papers can be great sources of stories, but understanding their limitations and how they came to be what they are is critical. "Trust but verify," as Ronald Reagan said. Or "distrust but verify," as Hillary Clinton did in 2015 describing the agreement to lift sanctions on Iran for its nuclear program.

4

Working With Statistics

Maggie Koerth

Maggie Koerth is the senior science writer at FiveThirtyEight.com. Although she's spent half a decade at a site best known for its political coverage, her work spans a broad range of subjects, from the sociology of gun violence to climate change and the sex lives of pandas. The common thread running through all these topics is her deep interest in how scientific research and data shape the choices made by people and societies—and how all those humans, in turn, shape the process of science. Maggie's work has won awards from the American Association for the Advancement of Science and the National Academies of Sciences, Engineering, and Medicine. She is a former Nieman Fellow and serves a member of the board of the Council for the Advancement of Science Writing.

Statistics and data analysis can be intimidating. I will absolutely cop to that. I have worked for more than 5 years at a publication dedicated to data journalism, and I still get nervous sometimes about using stats and data correctly in my stories.

I'm lucky enough to have a quantitative editor at FiveThirtyEight. When I experience a statistically significant amount of anxiety, I can take my fears to her and make sure I'm doing this stuff right. Unfortunately, not everyone is so lucky. But here's the good news. In the end, success with stats—for most science reporters—is less about whether you can memorize 5,000 detailed math tricks and more about whether you can remember to be suspicious of numbers.

Here is what I have learned in 5 years of working at a publication dedicated to data analysis and statistical inquiry: Data aren't facts, not an objective ones, anyway. Just because you have numbers doesn't mean you have truth. Data, instead, are a perspective, and you should treat them the same way you'd treat statements in an interview with a source. The best tips and tricks for working with data are things that help you interrogate numbers the same way you'd question a living, breathing source.

Maggie Koerth, *Working With Statistics* In: *A Tactical Guide to Science Journalism*. Edited by: Deborah Blum, Ashley Smart, and Tom Zeller Jr., Oxford University Press. © Oxford University Press 2022. DOI: 10.1093/oso/9780197551509.003.0005

Read the Fine Print

Numbers are not dropped upon us by the gods. They're created by people. Human scientists and researchers are the ones who decide what questions to ask, what the likely answers to look for are, what data might provide answers, how to go about collecting those data, and how best to interpret the results. At every step, there are choices, different ways things could be done. Those choices are not necessarily good or bad (although sometimes they are definitely better or worse). But each choice affects whether the number the computer spits out at the end is getting us closer to true understanding or just making everything even more muddled than it already was.

Because of this, one of the most important things you can do any time you are presented with statistics is to ask where they came from. How were these data collected? How do the people who collected them define important terms (especially ones you think you know the definition of instinctively)? Has the way the data were selected changed over time?

This is vital data analysis. To get the answers, you'll have to dig into the Methods section and Appendices of published research papers; request the data dictionaries and documentation files for databases; and actually read the fine print in government reports. What you learn can change your understanding of what the statistics you're looking at really mean.

Several years ago at FiveThirtyEight, we set out to do a story on the economics of farming. We'd seen data suggesting that the number of American farms—all 2 million of them—had basically stayed the same between 1982 and 2016. That was a surprise because experts and activists had spent those same 30-odd years warning about farm consolidation: small farms being gobbled up by a few, ever larger, estates.

So my editor sent me digging. We needed to know not only how many farms there were in 1982 and in 2016, but also how those farms were being counted, how the distribution of farm sizes had changed, and even what the definition of "farm" was. We found that the largest farms really *had* been getting larger. Farms over 1,000 acres still represented a small portion of all farms, but they now accounted for more than half of all cropland. Meanwhile, the number of medium-sized farms had fallen significantly. That was hidden in the big-picture count because of the apparent increase in very small farms—a category that had been defined in such a way as to end up including a lot of things that were really just exurban houses with very big yards.

The top-line numbers were accurate, but they obscured the truth. To write about these statistics well, you had to know what the numbers really meant.

Ask "Dumb" Questions About Risk and Probability

Handling everyday statistics well is mostly about critical thinking, asking questions, and digging deeper to find the truth underneath what you're told. In other words, it's just regular old journalism skills. The trick is remembering to apply them to numbers as well as words. To remember that the number is not the end of the story, but the beginning.

There are tips you can use that will help you do this better, but even then, it's less about doing arithmetic and more about remembering where the weak spots are—and poking them. Measurements of risk and probability, for example, are things that should always be poked at. That's because neither you nor your readers have a great internal sense of what a given risk means for your life. Frankly, no one does. Study after study has shown that the human brain is generally a pretty poor judge of risk and few of us are taught enough about statistics to overcome that innate failure.

A case in point is the risks of building a house in a 100-year floodplain. Quite reasonably, the average homeowner could read those words and come away believing that the chances of their property flooding during their own lifetime is small. It *sounds* like a way of saying the area will only flood once every 100 years.

But that's not what it means. And the best way to make sure you don't miss that nuance is to ask your sources what a term like that means. How much risk is that, exactly? Even if you think you know the answer, ask.

And then keep asking. Because what a "100-year floodplain" actually means is a place that has a 1 percent chance of flooding every year. Now that might also sound small, but that's why you ask, again, what that means. If you do, you'll learn that those years of risk don't happen independently of one another. If you own a home for 30 years, the chance of a flood happening during that time is actually 26 percent.

But even that doesn't always make the real risks clear because most people understand risk better as concrete numbers or examples, not percentages. So ask sources to frame risk for you this way and tell your readers that the chances of their house flooding (at least once in the life of a 30-year mortgage) are actually greater than 1 in 4.

You can see how that conveys something entirely different than the phrase "100-year floodplain" does. So ask the "dumb" questions. Ask experts to define buzzword terminology, even if (especially if) everyone around you is dropping it into their stories as though it needs no explanation.

When you're writing about risk, think about how small everyday risks might accumulate over time—and ask your sources about it. Pay attention to the details of what the risk is describing: the time period or number of events it covers; how a risk of a scary thing compares to the probability of other kinds of dangerous events we're more inured to; and how many actual events the percentage risk represents.

These are the questions that can make the difference between a misleading statistic and one that helps your readers make sense of the world.

Remember the Margin of Error

Uncertainty is a part of life and is almost certainly part of any statistics you may report. To do your job well, you need to know how big that uncertainty is and how it affects the way your readers might interpret those data in their own lives.

In the case of the 100-year floodplain, it turned out that the very idea of identifying floodplain boundaries is wildly inexact. Those seemingly confident maps are built on estimates. The uncertainty comes from a lot of different places, like the stream gauges that scientists use to collect data on the volume of water moving down a stream or river. These tools help us understand how often waterways are hitting flood stage. But many of these stream gauges have been collecting data for fewer than 50 years, so we don't know how well the information they've supplied represents what's normal. Meanwhile, while there are lots of stream gauges out there, there are also lots of places not getting measured at all. And that includes areas that have had documented cases of flooding.

The smaller the number of items in a sample—and the fewer the number of years for comparison—the less certain you can be about what your system of measurement is telling you. This is a reality that affects everything from opinion polls to medical studies and floodplain maps. Simply put, less information means less certainty, and more information means more certainty. But even when there is a *lot* of information, there's still some uncertainty baked in—and some possibility that reality will deviate from the estimates in the study.

For floodplains, for example, the maps aren't as exact as they can seem to homeowners. And flooding can very easily occur on the "safe" side of a floodplain boundary. If a catastrophic flood for the area is estimated to be 10 feet deep, the real thing could be 7 feet . . . or 13. All too often, the numbers we get in statistics are really ranges, not an exact point. When you're thinking about risk to your home and property, that is a pretty meaningful spread.

This problem pops up in all kinds of stories. As Laura Bronner, my former quantitative editor at FiveThirtyEight, likes to say: "No politician is ever 1

point ahead in a poll." That's because the margin of error (the range of uncertainty built into what the number actually represents) is larger than 1 percentage point.

There's no perfect way to do a poll. They'll all have some amount of error
baked in, depending on factors like how many people were polled, the framing
of the questions, or how well the people polled match to the general population in terms of demographics. A candidate who appears to be 1 point ahead
might actually be 3 points behind, allowing for margin of error. You can know
the error range—each poll has one. But you can't know exactly what the answer is within that margin. It's important to communicate poll results in a way
that won't make the polls seem more certain than they truly are.

Keep a Cheat Sheet

While there are an infinite number of ways to complicate your statistical
analysis, you are likely to find yourself running into the same questions and
problems over and over and over. As time goes on, that will make them easier
to remember. It also means a fairly small list of terms pinned above your desk
will help you catch the vast majority of issues.

What, specifically, you'll need on your cheat sheet will vary depending on
the topics you cover most often, but here are some good basics:

Sample Size: The smaller the number of people being studied, sites measured, or whatever else, the more likely it is that the results aren't gen-
   eralizable. Scientists often refer to the number of objects in a research
   sample as n If your $n = 1$, that's just an anecdote, not data. You don't
   know how well that person or thing represents the norm. This doesn't
   mean all studies with small ns should be completely dismissed: Some
   have valuable things to tell us. But it's crucial to know what the sample
   size is, how big that is relative to other studies that cover the same topics,
   and how the objects in the sample were selected.

p Value: This is basically a measure of how likely it is that a scientist's hypothesis is correct. Think of it this way: When you do an experiment, you're
   testing a hypothesis (if I poke this thing, it will flinch) against a "null hypothesis" (if I poke this thing, nothing will happen). If the data that are
   collected in the experiment tend to support the null hypothesis, then the
   p value will be higher. If the data tend to support your hypothesis, the
   p value will be lower. Generally, scientists accept that a study supports
   the hypothesis if the p value is .05 or lower, and that's when people will

start throwing around the words "statistically significant." But beware! Researchers can manipulate data collection to produce the p values they want, and factors like sampling bias (i.e., deliberate selection of test subjects most likely to support the hypothesis) still matter a lot.

Average/Mean/Median/Mode: I find it easiest to think about mean, median, and mode as different kinds of averages. They represent different things. Mean is what you're most used to thinking about when someone says "average." It's what you get when you add all the numbers in a set and then divide by how many numbers there were in that set. If five people work for a company and they make $17,000, $17,000, $20,000, $30,000 and $1,000,000, respectively, the mean salary will be $216,800. But you can see how the story changes if you report the median salary—the number in the middle. That's $20,000. The story changes again if you report the mode: the number that recurs most frequently. That's $17,000. Sources can use different kinds of averages to frame their narratives in different ways, so it's important to know what's really going on.

Absolute and Relative Risk: This distinction comes up a lot in healthcare reporting, when press releases will often tout a new treatment as reducing risk of a disease by some huge amount. That can be misleading because those statistics aren't usually telling you much about absolute risk: how likely someone was to get the disease to begin with. Instead, they're usually describing changes in relative risk: how the risk in the treated population differed compared to the risk in the untreated population. The absolute risk can be really small even if the relative risk is huge. Getting them mixed up or using them interchangeably can make your readers think a treatment is more effective than it really is or, on the flip side, that an activity is more dangerous than it really is.

Percentage Versus Percentage Point: There is an easy way to remember this difference: Percentage points are what happen when you're just counting within percentages. A 5 percentage point increase from 20 percent brings you to 25 percent. A percentage increase, on the other hand, requires a little more complicated math. To figure out what a 5 percent increase from 20 percent would be, you need to first calculate what 5 percent of 20 is, then add that result to your original percentage. The result is 21 percent.

Rates and Comparisons: As with absolute and relative risk, there's a difference between the absolute number of times a thing has happened, the per capita rate of it happening in a population, and the percentage change in how often it happens from one year to the next. This comes up a lot in crime statistics, where percentage comparisons between different years can be used to make it look like crime is skyrocketing when the absolute numbers (and the number of times the crime happens per person) are still fairly low.

The Takeaway

The more you practice being skeptical of numbers, the better you'll get at it. There are also lots of helpful resources. I recommend *The Curious Journalist's Guide to Data* from the Tow Center for Digital Journalism, a book that is available online; the Society of Professional Journalists' online Journalist's Toolbox site, which has a whole section dedicated to your math-related needs, including multiple statistics crib sheets; *A Guide to Statistics for Journalists*, a video webinar available from the Reuters Institute for the Study of Journalism in the United Kingdom; and the amazing in-person data journalism boot camps offered by Investigative Reporters and Editors (IRE).

But the most important thing to know is that numbers can—and should—be questioned. If you can remember that, you'll be well on your way to solid statistical reporting. That's the big rule. Everything else is just helping you narrow down what questions to ask—and asking them.

5

Fact Checking

Brooke Borel

Brooke Borel is a journalist specializing in science and technology. She's the ar-ticles editor at Undark Magazine *and has also written for* Popular Science, BuzzFeed News, *the* Atlantic, Scientific American, FiveThirtyEight, *and others. The Alicia Patterson Foundation, the Alfred P. Sloan Foundation, and the Gordon and Betty Moore Foundation have funded her work. She teaches writing at the Arthur L. Carter Journalism Institute at New York University and speaks on journalism and fact checking both nationally and internation-ally. Her writing has been anthologized in* What Future *and honored by the American Society of Journalists and Authors and the National Academies Communications Awards. Her books are* Infested: How the Bed Bug Infiltrated Our Bedrooms and Took Over the World *and* The Chicago Guide to Fact-Checking, *both from the University of Chicago Press.*

For a journalist, there are few worse feelings than the remorse over having to print a correction. All that hard work finding sources, learning about new—and often complex—topics, putting words to the page, and responding to a slew of questions and comments from your editor, and you still didn't get it right. Even worse: Your name is in that byline. You're the person who's going to get the upset email from a source who trusted you or a snarky comment from a cantankerous scientist with a huge following on social media.

You've also probably blown your chance at using that piece to highlight your best work for a fellowship, award, or job application. And you may have lost the trust of the person to whom you actually owe the most: the reader. After all, if the point of journalism is to inform the public on topics that may be of concern or interest to them, journalists do a great disservice when that information is inaccurate. This is especially true in science reporting, which covers the world's greatest challenges—from pandemics to climate change, often in the face of targeted disinformation campaigns.

To help prevent embarrassing missteps, many outlets and individual reporters have a system in place: the fact check. Fact checking is more than

Brooke Borel, *Fact Checking* In: *A Tactical Guide to Science Journalism.* Edited by: Deborah Blum, Ashley Smart, and Tom Zeller Jr., Oxford University Press. © Oxford University Press 2022. DOI: 10.1093/oso/9780197551509.003.0006 simply doing your due diligence while reporting, although that's important, too. Rather, fact checking is a step in the editorial process that can help catch errors that may have slipped in during every stage that a story goes through, from reporting to editing to rewriting to copyediting to layout. And it's a skill that pulls from many editorial jobs.

"I like to say that a fact-checker is a reporter who edits," says Tekendra Parmar, a former editor at *Rest of World*, who oversaw the publication's fact-checking program, and current tech editor at *Insider*. "So all of those skills that make a good reporter—the curiosity, the doggedness, the thirst for knowledge, whatever—those make a great checker." But what makes the difference, he adds, is also "having a great bedside manner."

Knowing how fact checking functions can help you to be more thorough in your reporting, as well as understand the larger editorial process, which could help you in all sorts of jobs, from fact-checking to writing and editing. Fact checking could protect you, or the outlets you write for, from legal trouble. And it may save you from the pain of adding a correction to the end of your lovely, hard-earned story.

What Is Fact Checking, Anyway?

This chapter focuses on editorial fact checking, which is part of the in-house editing process that takes place at an outlet *before* a story is published. (Another type, called political fact checking, happens *after* information is already out in the public, although it focuses mainly on speeches, debates, and statements made by politicians or other public figures. While many journalists participate in political fact checking—including some that is science focused at places like Snopes and SciCheck—it isn't relevant to our discussions here.)

In general, editorial fact checking takes place after the editor and journalist have a story in more or less its final form. In other words, they don't plan to do any major structural surgery. Then, someone goes through the story line by line, checking each fact against the original sources—and perhaps digging up new sources if the originals are found lacking or suspect.

Who that someone is may vary. Historically at many print magazines, and increasingly at many digital publications, there are dedicated fact checkers—people who had no part in producing the story and who are solely interested in making sure all the facts hold up (the "magazine model"). At most newspapers, there isn't much time to have yet another person get involved in a story, especially for breaking news or other timely items. Here, the journalist is typically responsible for the facts and may go back and review their work before publication; copyeditors may also help spot-check basic facts (the "newspaper model"). And, increasingly, there are publications that do a little bit of both, using the magazine model for longer or more complex stories and the newspaper model for breaking news or short items (the "hybrid model").

For more on these categories, as well as insight into which science publications use them, check out "The State of Fact Checking in Science Journalism," a white paper (available for free online) I wrote with a team at the Knight Science Journalism program at the Massachusetts Institute of Technology in 2018.

Any of these fact-checking models may also be found in radio or video, where the work generally, though not always, happens in scripts before the final product is made. The basic work remains the same. "A fact is a fact. That fundamentally doesn't change across different mediums," says Wudan Yan, an independent journalist who has fact-checked for both print and podcasts—though she added that the presentation of facts naturally varies. In print, for instance, the facts may be a little more obvious, Yan says, while in audio, a character often details a sequence of events. That character may not be entirely reliable, she adds, so "you have to triangulate like you would do in an investigative story."

No matter the model or the medium, the key is to have someone read a story with only the facts in mind. This person shouldn't get hung up on structure, storytelling, or line edits—unless for some reason the very framing of a story may mislead readers or obscure the facts. Rather, the fact checker focuses on three key questions: Do they have sourcing for every discrete fact, statement, or claim? Is that sourcing trustworthy? And, do all of the facts add up to a story that is true and complete?

How Does It Work?

The details of fact checking may look a little different from one publication to the next, but most publications follow the same broad process. Still, it's a good idea to confirm the details with individual outlets. "It's the publication or the podcast or the digital company you're working for that makes the difference, because everyone has their own requirements for what they want fact-check-wise," says Eva Dasher of Dasher Editorial, whose clients include *National Geographic*, *Scientific American*, and *Science Vs*.

In the magazine model, the journalist annotates every fact in their story, noting the source from which it came. Usually, this means making a draft of the story in Microsoft Word or Google Docs and including the source information as either comments or footnotes.

Next, the journalist organizes the sources, which may include interview recordings or transcripts, contact information for sources, handwritten notes, scientific journal articles, photographs of locations mentioned in the piece, and so on. For electronic files, the journalist will gather these materials in an email or a file-sharing service such as Dropbox; for hard copies, they may scan relevant pages or, if the item includes many pages from a particularly long text, they may send the whole book or document to the fact checker and note the relevant page numbers. Stories with highly sensitive sources (i.e., whistleblowers and other individuals whose identity must remain confidential), the journalist might pass the information via an encrypted messaging service or on a computer drive or laptop.

Now the fact checker gets to work. Some prefer scratching comments on hard copies, in which case they may print the story and cross out each fact after they confirm it, writing potential corrections in the margins. Others may work directly in the electronic file, for instance striking through text and writing notes as comments or footnotes. It doesn't matter, so long as the work is methodical, consistent, and includes these steps: identify a fact, check it against the source, vet the source, and repeat.

Fact checkers who are working with science stories will also keep a close eye on how that science is described, confirming whether a metaphor is apt or whether a layperson term is an accurate substitution for a bit of jargon. Sometimes fact checkers may even re-interview the people who appear in a story, double-checking information with them by phone or email. (Take care with this: Most outlets don't allow unpublished stories, or even verbatim quotes or excerpts from a story, to be shared directly with sources.)

And every once in a while, the fact checker should take a big step back and ask: Based on the sources, does this story do justice to the truth: Does everything hang together in a way that makes sense? And, is there anything *missing* from this story that, should it be included, would make it more accurate?

In the newspaper model, journalists may go through their own stories and double-check their work. In some ways, this is easier since it doesn't usually require a detailed annotation or sharing a long list of sources with someone else. But it's also very hard to take a critical look at your writing and admit when a source isn't quite up to snuff or a turn of phrase isn't quite right. Here, it is vital for the journalist to read the piece as though they didn't write it, perhaps considering it from the point of view of one of their strongest critics. And copyeditors—at least at publications that still employ them—often help check some facts in the newspaper model, like the spelling of sources' names, basic geographical information, and common typos (should that be *billions* or *millions*?).

In the hybrid model, the editorial staff divvies up which stories get which treatment. Usually, the magazine model is used on longer or more complex stories. These stories may have gone through many rounds of edits or have many, many facts within them, which raises the chances that an error snuck in. Publishers will have already invested heavily in such stories, which may involve many months of reporting, special art treatments or layouts, and so on. An error in a story like this, especially if it is a bad error, can make all that effort worthless, which is why some outlets are willing to put in the extra work to make sure the story is airtight. With the hybrid model, shorter pieces that are more straightforward and easier to check—or pieces that are very time sensitive—usually go through a newspaper-style check.

Why Should I Fact-Check?

The legal answer to why you should fact-check is: So you don't get sued. I'm not a media lawyer, so if you have specific questions on the legal implications, you should go to an expert for answers. But the types of things that fact checking *may* save you or your outlet from include libel, invasion of privacy, copyright infringement, and—though it isn't illegal—plagiarism.

The personal answer for this question is: Your reputation as a journalist matters. It's really all you have. It will help you get work; it will help you gain readers' trust. If you have a lot of sloppy mistakes in your writing it will potentially hurt your career. Sure, there are journalists who have made grave mistakes but who have risen to a level of fame where it doesn't seem to matter. But those journalists don't exactly give the profession a good name. And do you really want to be one of *those* journalists?

The philosophical answer is: Readers deserve accurate information. As journalists—or as any writers of nonfiction—we have an implicit contract with the reader that what we are telling them is true. We are implying that we did the work to confirm each and every fact in the piece. We're supposed to be serving the public, so we should make sure we are serving them well.

But What if I Don't Work With a Fact Checker?!

Not all outlets use fact checkers. Some may not even force you to double-check your own work before you publish, which means it'll be up to you to remain disciplined. Do not fear: There are several steps that you can take to make sure your stories are accurate.

First, create a system and stick with it. For instance, you might build some time into your deadline so that you can step away from your story—whether it's for a day or just a 5-minute coffee break—and come back to it with fresh eyes. Find ways to trick yourself into thinking you're reading it for the first time by printing it out in a new font and moving to a new room to read it. Underline each fact in the piece and ask: How do I know this is true? And don't be easy on yourself. You should read the piece with a critical eye, as though you are trying to pick apart someone else's writing.

Second, stay organized. Be sure to note any source that you use as you are writing, so you know how to find it again later (i.e., start a separate document where you include every quote and the timestamp where it appears in your audio recordings and another one that lists every scientific article you're relying on plus the relevant page number). Clearly label the file names for all of your sources and keep them organized on your computer. You should be able to retrace your steps, both when you are fact-checking your story and even months or years later, should someone try to challenge it.

And finally, if you don't have the time to go back to every single source thanks to a tight deadline, at the very least double-check the big claims and make sure you have solid sourcing. For instance, if you're accusing a scientist of misconduct or a major tech company of misdeeds, go back to your original documentation and make sure it supports the claims—and make sure that you have hard evidence that you attempted to get a response from the accused, to get their side of things in the original piece. And on the other end of the spectrum, check the very small facts that are both easy to get wrong and easy to look up: the spelling of names, university affiliations, dates, geographical locations, and so on. When you get the small stuff wrong, readers notice.

The Takeaway

Fact checking is a key editorial step that will help ensure that your stories are accurate. Some outlets have dedicated staff or freelancers who do this work; others require the journalist to fact-check their own stories. Either way, you should always be careful with your sourcing and stay organized.

Understanding how fact checking works can improve your reporting skills, as well as help you to prepare your own stories for outlets that do an old-school magazine fact check. And editors who know how fact checking works can help ease their writers through it and also flag potential problems for both the writer and the fact checker.

Freelance fact-checking can also be a good gig; as of 2018, the Knight Science Journalism program at the Massachusetts Institute of Technology white paper on fact checking found that fact checkers make an average of about $30 per hour, and some can bring in much higher rates. Many journalists, and in particular freelancers, have side gigs as fact checkers. Others build entire businesses based on fact checking. Dasher is one of them. "Fact-checking has always kind of been my bread and butter. I always know it's there," she says. "And there's a great need for fact-checkers. I've always thought they were going to go away, and they really haven't. I think they've become more important."

PART II
THE CRAFT OF STORYTELLING

6

A Foundation in News

Alicia Chang

Alicia Chang is an editor at the Associated Press and specializes in health, science, and the environment. She previously served as the AP's Los Angeles-based science correspondent for more than a decade. She spent the 2015–2016 academic year as a fellow in the Knight Science Journalism program at Massachusetts Institute of Technology and was a 2018 entrepreneurial journalism fellow at the Craig Newmark Graduate School of Journalism at the City University of New York (CUNY), where she is an adjunct assistant professor. A native of Guayaquil, Ecuador, she grew up in the United States and earned a bachelor's degree in chemistry from Queens College CUNY. When she's not editing or teaching, you can find her boxing.

I once had to cover a space shuttle parade on a bicycle.

After 25 missions, the baby of NASA's shuttle fleet retired in Hollywood style with a plodding journey through city streets from LAX (Los Angeles International Airport) to a hangar on the grounds of a science museum, where it would spend its twilight years on display.

This wasn't your typical parade story. The logistics were as complex as an Einstein equation, the stakes enormous. After all, *Endeavour* logged 123 million miles in space. What could possibly go wrong with its final mission—its terrestrial homecoming?

Even before *Endeavour* moved an inch, I spent days scheming: I drove the length of the 12-mile route several times, chatting with residents, memorizing landmarks, and noting potential choke points. I prepared a draft in advance with context and background that could be topped with reaction, progress, and setbacks. I hatched a filing plan with the editor on duty to dictate quotes or email publishable updates.

On the big day, I hopped on my bike and tailed *Endeavour* as it crawled through neighborhoods atop a monster 160-wheel carrier under the gaze of heavy security and constellations of spectators. Preparation is key when covering spot news, or news happening in the moment. And so is flexibility,

Alicia Chang, *A Foundation in News* In: *A Tactical Guide to Science Journalism.* Edited by: Deborah Blum, Ashley Smart, and Tom Zeller Jr., Oxford University Press. © Oxford University Press 2022. DOI: 10.1093/oso/9780197551509.003.0007 because there are always surprises. I clocked in for another day of coverage when the move fell way behind schedule, eventually arriving at the destination 17 hours late.

Deadline journalism is thrilling and rewardingly difficult. You're an eyewitness to history-making events and milestones. You also juggle your journalistic powers simultaneously: hitting the accuracy target, explaining clearly and concisely, and presenting in an engaging way, all under a time crunch.

The daily science story—whether it's read, heard, or seen—informs, excites, and illuminates. Ideally, it grabs people who would otherwise not care to pause and pay attention—and inspires them to share the story at the dinner table, around a virtual water cooler, or on their social network.

Science as News

Compared with other beats, there is usually less breaking news in science journalism that requires pushing out articles on the spot.

An exception is the news conference. During *Endeavour*'s achingly slow road to retirement, there were media updates on wins and hiccups. Often, press conferences will be called after a breaking event, such as a rocket launch or a spacecraft touchdown; to coincide with the release of a major report, such as by the Intergovernmental Panel on Climate Change; or as arranged by a journal to highlight a study it deems important. Since there's no exclusivity, the goal is to boil down the essence and pluck out the most colorful quotes in a speedy way.

If a discovery is publicized through a news conference—and hasn't yet been published—this should raise a huge red flag. History is littered with "science by press conference" that was later debunked: a pair of chemists who claimed to have achieved cold fusion, a group founded by a religious sect that claimed it cloned humans, or a biologist-activist who claimed genetically modified corn causes cancer in rats.

For the most part, the pursuit of discovery is slow, driven by observations and trial and error. Chances are that story you saw about how human ancestors evolved or the newest advance in brain science was promoted by a journal. The main peer-reviewed ones often are embargoed, affording journalists a sneak peek of the research a few days before publication to allow for independent vetting and careful explanation. Was the hypothesis rigorously tested? Does the conclusion make sense? How does this expand our knowledge? What are other possible explanations?

The Covid-19 pandemic speeded up research, casting aside this traditional system. In a public health emergency, updates about vaccine and treatment studies and genetic research into the coronavirus can't wait to pass through the regular funnel and instead are shared in real time among researchers and the public, through either online preprint servers or company press releases.

When deciding what to cover—whether you have no time at all or just a few days—use critical thinking: Quiz the claims. Weigh the evidence. Research what came before. Tap outside voices for perspective. It's critical to bring healthy skepticism to extraordinary claims whether they're published in a bona fide journal or posted to an online site serving as a first draft of science. It's just as important to explain what is known and what knowledge is still lacking. Misinformation thrives in a vacuum of information, and being up front plugs the void.

Medical and scientific conferences are another venue where results are relayed. Like preprints, abstracts, or summaries of research, they aren't reviewed by outside experts and the same caution applies. Be sure to attend presentations and listen for barbed questions from the audience.

Sometimes science is truly spontaneous, like a sudden earthquake or the next outbreak hot spot. Even one of the most pressing global issues of our time—climate change—can figure as news as a hotter world intensifies dangers like wildfires, hurricanes, and other extreme weather-related events.

Though spot news stories may feel hurried, they are essential to informing the public. The constant drumbeat of science reminds us that there is always wonder in our daily lives.

Beating Deadlines

The beauty of science journalism is the diversity of topics. One moment, you could be covering the latest black hole discovery and then switch gears to the latest thinking on animal cognition. This bottomless buffet of options can make reporting on short deadlines challenging: How do you become an instant expert no matter your medium?

A trick: Invest in spadework off deadline. Identify several core subjects you'll likely cover again and again. And then make cheat sheets.

When I was a reporter in Los Angeles, earthquakes lurked as a constant threat. Unlike other disasters, earthquakes can't be predicted, and when the ground heaves, you're racing to be first with the news. Besides signing up for text messages to be alerted (even while asleep) of seismic shaking anywhere in the world, I kept files of background in ready-to-publish form (the death toll of the last major quake, simple explanations of plate tectonics, historical nuggets for context) to be pulled out and plopped into a breaking story at a moment's notice.

Even some breaking news events can be planned in advance. When I was on the cosmos beat, I borrowed a lesson from obituary writing. Many news organizations have prepared stories about the lives and achievements of notable people long before their deaths so when the eventuality comes they can move the obit out fast. Whenever there was a Mars mission or other robotic space quest, I prepared three versions of a story in anticipation of success, failure, or unknown outcome. It may sound extreme (one might even say macabre), but it eased the pressure of hunting for the perfect words on the spot.

Advanced legwork is particularly helpful for stories that move at an incremental pace, like climate change. In times of lull, squirreling away key facts and figures can help give you a slight edge when news happens.

Besides having background at the ready, cultivate sources who can comment on little notice. By forging relationships outside the pressures of a deadline, the odds are greater of reaching the people you need on deadline. It's important to make broad and wide contact so the same experts aren't overused and to ensure representation of underrepresented groups. Ask yourself: Whose voice is missing?

Contact experts early in a breaking situation since it can take several rejections before you land an interview. Scientists aren't fluent in every topic, so make sure the expertise matches. For example, don't seek a cancer specialist for comment about the ins and outs of a vaccine study.

Beware of false balance and don't give a megaphone to viewpoints when the evidence is heavily and convincingly tilted in one direction. Science isn't always definitive, and there are legitimate times when alternative explanations should be raised. But, when there's broad consensus on a matter such as smoking causes cancer or human-caused climate change is real, there's no room for the "other side."

Sometimes there's no choice but to start from zero on a completely unfamiliar topic. How do you ramp up? Many disciplines have professional societies, such as the American Statistical Association, American Geophysical Union, or their international counterparts, so tap them first for sources and branch out from there.

Even if you've never covered a topic, there are all-purpose warm-up questions: What's new here? Who's affected? Why should people care? What's the evidence? Who funded this? Who else should I contact? What else am I missing?

Think of a breaking story as a layer cake. Don't fret about pushing out a fully baked story from the go. Start off short, just 150 words leading with what you know will lurch you ahead of the competition. In the next take, sprinkle in quotations and background. The story will grow and deepen with more reporting. The first draft isn't always elegant, but you'll have the chance to revise with each version. Breaking it down into chunks makes pumping out stories with the seconds ticking less daunting—you will beat and not just meet deadlines.

Beyond Breaking News

It's a rush to chase news—to report when an outbreak crosses into a pandemic, when a fossil find shatters our assumptions, or when Einstein's general theory of relativity is confirmed for the millionth time.

But, the follow-up story is often overlooked. Whether to keep pursuing a thread can depend on competing news demands, audience appetite, and if there's room to explain or dive deeper.

Take a report by the United Nations detailing how countries' pledges to slash greenhouse gas emissions fall short to avoid a climate disaster. It's news worth sharing as it helps us understand the stakes of political inaction. Since it's far-fetched that any nation would automatically cancel coal plants or shake the status quo in any significant way as a result of the report's release, there's probably no need to spin it forward for the next day (although you can revisit down the road to see if promises were kept). Similarly, stories about the unearthing of dinosaurs or the evolutionary advantages of glow-in-the-dark critters are always fun reads, unless there's access to fieldwork, it's likely one and done.

Continuing news events like the Flint water crisis, the Fukushima nuclear meltdown, or a journey to another world bear following up to see what unfolds. In the case of a health or environmental emergency: How are the affected faring? What's the fallout? For space exploration: What's the first chore at the destination? How's the human team managing?

If the goal of spot news is to serve up facts, imagine the follow-up as a chance to add context and explain in a fuller way, to tell the story behind the story, to break from pack reporting. In an era of commodity news, an update—whether it's the next day or days later—can be a way to surface original reportage.

When NASA's *Curiosity* rover landed in the summer of 2012, I learned from a source that one of the flight directors planned to go on "Mars time" with his family. Since days on Mars last longer, every mission involves a dedicated army of scientists and engineers temporarily synching their body clock to the red planet, which leads to wacky work, sleep, and meal schedules. It's unheard of for an entire family to join the experiment and live this way. I knew it would be a story everyone would talk about after a successful landing, and my instincts proved right.

The pandemic declaration in March 2020 offers a case study of follow-up journalism to the extreme. When the World Health Organization finally said the "P" word, it became the singular story that news organizations pursued for months and still cover to a great degree today. The arc of the coverage showed there were many entry points for spot news: a new hot spot, record cases or deaths, pauses and reopenings, vaccine progress, and the status of vaccination drives. The quest for scientific understanding ran throughout the coverage, but with practically every corner of the globe touched in ways big and small, it morphed from a science story to a societal one.

A follow-up doesn't always have to be another story. It can take different forms, such as an explainer, Q & A, data visualization, or other visual medium. However it's packaged, it's important to include several notes of background to help audiences follow along.

The Checklist

For quick-turnaround stories, set an internal deadline for reporting so there's time to put your copy under a microscope before filing. You are your own backstop.

Spellcheck names and confirm titles. Quadruple check numbers. Inspect for context and strike out overgeneralizations. When you make a mistake, correct the record as soon as possible.

Truth is our North Star. Hype about a "first" or "breakthrough"—or worse, inaccurate framing—only sows public harm and fuels mistrust about how news is gathered and how science is tested.

With fast-moving stories, there's less time to pore over word choice or fiddle with a turn of phrase, but crisp writing doesn't have to be lost. Your prose can still sparkle. Circle verbs and replace passive words with active verbs. Break down a complex concept with an analogy, a comparison to a friendlier, more everyday idea. For example, cells are like factories, with different departments performing different tasks to work together.

Writing clearly means eliminating jargon—the technical language that scientists use to communicate with one another. It's not about dumbing it down; rather, it's about making science more accessible and reaching broader audiences by decoding "science speak." Instead of anthropogenic, try human-caused. Substitute spread for transmission. Rather than clinical trials, use studies, experiments, or tests.

Journalists are a competitive bunch, striving to be first and right. It's possible to be both. But if there's ever any doubt about accuracy, slow down and report it out until you're confident it's airtight. With so much deliberate false information swirling, let's do our part and not add to the noise.

The Takeaway

Distilling science's complexities and caveats on a tight deadline is a skill that can be perfected. It's our commitment to report the day's news, whether it's a discovery or a disaster, with rigor, precision, and compassion.

We must be up front with what we know and, more importantly, what we don't know and still need to learn. You need not be an expert in everything, but you should question everything. Even with the clock ticking, stories should carry different voices and be representative of society.

Deadline reporting is truly a public service. Now go do it and flex your superpower.

7

Story Structure

Deborah Blum

Deborah Blum, a Pulitzer Prize–winning journalist and author, is director of the Knight Science Journalism (KSJ) Program at the Massachusetts Institute of Technology and publisher of Undark *Magazine. She began her science writing career at* The Sacramento Bee, *where she won the Pulitzer in 1992 for a series on primate research. That series became her first book,* The Monkey Wars. *She has published five books since, most recently* The Poison Squad, *a 2018* New York Times Notable Book. *She has written for newspapers ranging from* The New York Times *to* The Guardian, *and magazines including* Wired, Time, Discover, Scientific American *and* National Geographic. *She was a journalism professor at the University of Wisconsin–Madison before becoming KSJ's fourth director in 2015.*

In the winter of 1933, at the bar of a shabby little speakeasy in the Bronx, a small gathering of down-on-their-luck New Yorkers plotted a murder. It would not turn out as they had hoped—at all. Perhaps if I tell you that the city newspapers later nicknamed the chosen victim "Mike the Durable" that will give you a hint about why.

The tale of Mike Malloy remains one of my favorite poison stories from the history of forensic toxicology, a subject I've been covering for years. I like every twist, every unexpected turn, and I also like sharing them. But before I do, let's pause for a moment. Let's save some of the good stuff for later. And let's do that not only because it's fun but also because this technique—holding the reader at a moment of suspense, foreshadowing the story—is a useful lesson in one of the more popular techniques used by narrative writers.

Good narrative writing is often as much technique as it is talent, sometimes more. The best narrative nonfiction writers often turn to time-honored tools of fiction writers for effect: plot and pacing, character and drama, and, yes, suspense. And they understand, to paraphrase the Pulitzer Prize–winning science writer Jon Franklin, that a good story just can't spiral out in all directions like a serving of spaghetti. The story needs form, shape, a structure designed

Deborah Blum, *Story Structure* In: *A Tactical Guide to Science Journalism.* Edited by: Deborah Blum, Ashley Smart, and Tom Zeller Jr., Oxford University Press. © Oxford University Press 2022. DOI: 10.1093/oso/9780197551509.003.0008 to pull the reader from start to finish. "The craftsmanship of the writer is no less beautiful than that of the cabinet maker or the builder of temples or fine violins," writes Franklin in his guide to the craft, *Writing for Story*.

Yes, this may sound grandiose, but the emphasis on craftsmanship is pure pragmatism: a knowledge of the basic structures that narrative science writers use to build an effective story. I think of this approach as journalistic architecture. Once a writer has the story blueprints in hand, so to speak, then he or she can decide which structure best fits the facts of the story—and where to slot them into place.

To continue the architecture analogy for just a minute, we're not talking about one-size-fits-all story plans: different stories often demand different shapes. My 90-year-old tale of a murder-resistant victim eventually defeated by poison, for instance, can't be told in a classic newspaper inverted pyramid style. Other forms—say a broken-line narrative structure—serve it better. I'm not going to review every possible structure here. But let's consider some of the options.

The Inverted Pyramid

Writers usually think of the basic inverted pyramid structure as beginner journalism but it still offers a handy step-by-step guide to writing a news story, one that enables you to write very quickly on deadline.

Why is it called an inverted pyramid? Picture the triangular shape of an Egyptian pyramid turned upside down so it balances on its peak: big at the top, tiny at the bottom. The key message for news writers is simple. You put the most important information at the top and the least important at the bottom. That way if a reader doesn't finish the story, you've at least delivered the good stuff up top.

Consider the first paragraph of Boyce Rensberger's *Washington Post* story about the catastrophic loss of a NASA space shuttle in 1986: "The space shuttle Challenger, carrying six astronauts and schoolteacher Christa McAuliffe, exploded in a burst of fire 74 seconds after liftoff yesterday, killing all seven aboard, and stunning a world made witness to the event by television." Even if you didn't continue on to the next paragraph—which I can guarantee you most people did—that sentence neatly delivers the key points in the story.

Fortunately for us, not all news concerns horrifying disasters. Of course, this also means that many news stories lack such dramatic power. But the inverted pyramid offers a useful guide to dealing with that as well. I think of it as a four-step plan:

1. The broad top represents the lead (or lede), basically the opening sentence, paragraph, or section. Its job is to engage the reader with the story. In the case of the exploding space shuttle, the facts serve that purpose. In the normal, nonexplosive world of standard news stories, the writer needs to be more creative and often less comprehensive to pull the reader in. As an example, let's start with this one-sentence lead from a 2021 *Washington Post* story on a mysterious brain disease spreading in eastern Canada: "Alier Marraro is stumped." Those four words are meant to intrigue only.

2. In the next step, the writer continues with the "so-what section," or what journalism insiders sometimes call "the nutgraf", which fills in more details, pulls the reader further in, and explains why the story is worth the readers' time. In this case, it's that the disease poses a yet uncontrolled and alarming threat. Marraro, a researcher at a New Brunswick medical center, is again quoted, saying: "The suffering is immense."

3. Then follows what I think of as the "guide section," which summarizes key issues in the story. Here that includes the history of the disease, the often-crippling symptoms, questions about its mysterious cause, and the public health response. These individual points are then explored in a logical sequence that sets up the story's ending.

4. The story ends with a "summary point"—yes, now we're at the bottom point of the inverted pyramid—which may also suggest a resolution. In newspapers, this is often a closing quotation. Again from Marraro: "Fear is understandable. But we are working for hope."

I admire the way this structure thus keeps a story in forward motion. But let's now consider other approaches.

The Diamond Structure

An editor at a popular magazine once told me he considered the diamond structure the most useful tool he knew. Like the inverted pyramid, it's a straightforward story shape, one more tailored to narrative or feature writing than to news.

By telling you that it's straightforward, I'm pretty much giving away the fact that we're not talking about a multifaceted gemstone of a structure. This is instead the diamond shape used on playing cards: straight sided, sharply pointed at the top, widening out in the middle, and narrowing back down to a sharp-pointed bottom.

Those who like geometry will recognize it as a type of rhombus. But what does a rhombus have to do with writing? Basically, it's a structure that tells you to start a story with a very sharp focus, let it broaden out to the big picture involved, and then bring the story back to the focused point of the beginning.

Usually, this means starting a story with a person or an anecdote about that person that illustrates the issue at hand. This allows the writer to then open the bigger questions revealed in that story, explore them, and then return to or finish the anecdote as the story comes to a close. It's a great way of making issues, which might seem abstract or academic completely, compellingly human.

One of my favorite examples is a 2010 story in the *New Yorker* by Atul Gawande, called *Letting Go*. In it, Gawande explores an ethical question: In its quest to prolong life at all costs, does medicine cause more suffering than it prevents? But he doesn't start with the issue itself; he starts with the story of a young woman, newly pregnant, who is suddenly diagnosed with a terminal cancer. She tells her doctor that she's willing to undertake any risky treatment so that she can survive to raise her child.

Many of us have known people in similar situations. I had a friend, an exceptionally talented writer in his early 40s, who was diagnosed with metastatic melanoma. In his determination to fight it, he enrolled in an experimental program for cancer patients who had been sent home to die by their oncologists. He lived at least 9 months longer than he might have otherwise, but the drugs used were so toxic that by the time he died he was blind, and his kidneys had failed. This is the conundrum that Gawande explores, starting with one desperate quest to fend off a lethal disease, then exploring others and their consequences, interviewing doctors, nurses, ethicists, and patients.

Through such interviews and experiences, he broadens the story from its single-patient start to an inflexion point where the reader reaches the central question of the story: Is life at all costs really life? Once he reaches that question, he starts narrowing the focus back until the story concludes with the heartbreaking end of the young woman's story, one that underlines the question with human emphasis.

The diamond structure can be both simple and powerful. And, it emphasizes another key point about good storytelling: It's always better to know the start and finish of the story before you start writing.

The Narrative Arc

You sometimes hear writers extolling the idea of a "perfect narrative arc." Again, that sounds fancier than it really is: It's just another way of structuring and planning a story.

It's also a structure that no one entirely agrees on. If you google "narrative arc" and look at the images, you'll find lumps, jagged lines, and simple curves, all with different instructions. For our purpose, I'm going use a curve, basically the arch of a rainbow as it rises, bends, and descends.

The arc of the story essentially follows that curve. It begins with a conflict, rises to a crisis point, and inclines back down toward a resolution. In *Writing for Story*, Franklin describes these three points—conflict, crisis, resolution— as the fundamentals of any good story. He also argues that knowing the path from conflict to resolution helps the writer find the story's primary point of view.

Consider Franklin's Pulitzer Prize–winning tale of a failed surgery, *Mrs. Kelly's Monster*. It's the story of a risky operation to remove a tangling mass of blood cells from a woman's brain, with the hope of ending years of pain and risk. Franklin planned to follow the arc of a courageous woman's willingness to risk all to defeat the enemy, a tale of classic conflict and resolution—but the operation killed her.

I thought, Franklin said, that as the patient was lost so was the story. But he then realized that he could change the perspective. So, *Mrs. Kelly's Monster* became instead the story of a neurosurgeon attempting a dangerous but heroic surgery (the conflict). The arc of it rises until the surgery starts to go wrong (the crisis), and follows it beat by heartbeat until the patient's death—and then just a little beyond it.

"The story is a classic of white knight and maiden, and in this case the white knight failed," Franklin said. "But the final thing is, he had to get up after that and go in and work on somebody else."

And that, I'll admit, might just be a perfect narrative arc.

Zipper and Braided Narratives

Zipper narratives are wonderfully versatile structures. You can put them to use in a long-form magazine story and use them to plan out a book as well.

The concept is based on the idea of the zippers we use in clothing, which use interlocking teeth to bind two separate pieces of fabric together. The zipper narrative essentially binds two separate but related stories into one interwoven tale.

For example, you've decided to tell the story of a scientist who has dedicated her life to exploring the link between Parkinson disease and toxic chemical exposures. One of your threads is the story of that scientist's lifelong quest to make that connection. The other might be the story of Parkinson disease itself, when it was discovered, how it affects the nervous disease, why it remains mysterious. You zip back and forth between the two story strands, figuring out when you allow one story thread to stand alone, when you zip them together, when you pull them back apart.

Of course, you can always weave together more than two story threads. A more intricate version is a braided narrative. Named for the traditional hair braid, this approach weaves three story strands together, and apart, and back together.

In this case, you might have another strand you want to add to the Parkinson's story—perhaps a story from your own history, a friend or relative stricken by the disease. You create a third story thread that follows the progression of their disease, adding a vivid human portrait of its effects. You may want this third thread because it strengthens the story. But again, the trick is figuring out when each tale stands alone, how long you spend with the different strands, and when you weave them together.

Both structures are built on elegant architecture. And, to repeat myself because it matters, both send the same message to the writer: plan and pace your story in advance.

Broken Line and Layered Narratives

Lee Gutkind, founder of the magazine *Creative Nonfiction*, once described the foundation of narrative writing as this: Story-information-story-information-story-information-story.

The writer begins with a seductive amount of story, then spools out some information, returns to the story before the reader gets bored, calculates when to add in the next packet of facts, and continues this layering until the end. If the writer is good enough, the reader is only aware of the story.

Some people also refer to this as a broken-line narrative. For this case study, let's use my story of a 1930s murder plot—and occasionally break the chronology of the tale to provide background, explanation, and exposition. Again, there's a calculation here: How many breaks from the main narrative, which the Delaware-based writer McKay Jenkins calls "side-trips," can you take before the reader loses interest in the main story and abandons the journey?

Let's test it here with that murder story I promised you.

In the winter of 1933, a group of five New Yorkers gathered at an illegal speakeasy in the Bronx to plot a murder. It was meant to make them rich, or at least gain them each some much-needed cash. Instead, it sent them all to the electric chair.

Factual background: The little bar is illegal because it operates during Prohibition, which made the sale of alcohol illegal. The group, which includes the bar owner, consists of men struggling for money during the Great Depression.

Back to the story: The five come up with a plan to buy life insurance policies for one of the bar's customers, an impoverished alcoholic named Mike Malloy. Once he's insured, the owner of the speakeasy starts serving Malloy free drinks of a poisonous version of alcohol.

Factual background: Methanol or wood alcohol had become something of a plague during the Prohibition period. It tastes like ethanol, regular drinking alcohol. But unlike ethanol, it metabolizes in the body into formaldehyde. As a result, it killed thousands during this period.

Back to the story: Malloy turns out to tolerate methanol unexpectedly well. The plotters try other methods, from glass in sandwiches to hiring a cab driver to run over him. Nothing succeeds. Finally, desperate, they turn to another poison, carbon monoxide.

Factual background: Carbon monoxide is a notably efficient killer, ruthlessly replacing oxygen in the bloodstream and causing a kind of chemical suffocation. It's also extremely easy to detect in a corpse.

Back to the story: When Malloy passes out from drinking, the plotters return him to his room, connect a rubber tube to a gas appliance, and put it in his mouth. He dies almost immediately. Within days, they move to cash in the insurance policies. But they've also been sharing their frustrations with friends. This mistake leads to a police investigation, a toxicology analysis, and multiple arrests.

As you'll see, there's a solid amount of toxicology layered through the many attempts to kill Mike Malloy. But the story itself is so peculiar that I can use its natural theater to pull the reader past any aversion they might have to chemistry and on to the unhappy end.

In case you wondered, that end would be the state-ordered electrocution of all the conspirators. A newspaper reporter of the time, describing the crack and sizzle of the electric chair, called it the "state's toast to Mike the Durable."

The Takeaway

Story structures are useful tools for narrative writers. This chapter gives examples of some of the most popular ones and how they work. They are especially valuable in preplanning a story and in thinking about how to best weave technical information into a tale. And they are a reminder that plotting a story in advance is one of the smartest things we do.

8

Audio Storytelling

Elana Gordon

Elana Gordon is a journalist and audio producer covering global health for The World, *a national radio program from* PRX *and* GBH *in Boston. In that role, she also hosts an online discussion and podcast series about the pandemic, in partnership with Harvard's T. H. Chan School of Public Health. She previously covered health in the Midwest for the* NPR *affiliate in Kansas City and was a founding member of* The Pulse, *a weekly health and science show from* WHYY *in Philadelphia. Elana's work has been featured in* Kaiser Health News, Undark, *and* The Washington Post, *as well as on* NPR *and the podcasts* 99% Invisible *and* Criminal *and her reporting has received multiple regional Edward R. Murrow and national Public Media Journalists Association Awards. She was a 2018–19 Knight Science Journalism fellow at MIT.*

In early 2020, I traveled to Lisbon for a reporting project on addiction. I had interviews lined up and arrangements to shadow those on the front lines. As a global health reporter for *The World* from PRX and GBH and a podcast producer, my goal was to file a radio feature and podcast episode about Portugal's approach to drugs, just as communities in the United States were considering similar decriminalization policies and wrestling with the evidence.

One of my first interviews was with someone I knew would be a central character in this story: Portugal's drug czar, João Goulão, considered a key architect in designing the country's policy overhaul 20 years ago. I knew he'd have stories to share. But how to draw them out in a way that works for radio?

We could have gone to his office, but thankfully Goulão agreed to meet me at a former drug market, abandoned now but full of the story's history. That morning, we walked through the overgrown paths, through Goulão's memories and reflections. I had my headphones on, microphone out, and recorder rolling to capture it all. I wound up with a more vivid conversation in an environment that created greater opportunities for discovery and observation.

Elana Gordon, *Audio Storytelling* In: *A Tactical Guide to Science Journalism.* Edited by: Deborah Blum, Ashley Smart, and Tom Zeller Jr., Oxford University Press. © Oxford University Press 2022. DOI: 10.1093/oso/9780197551509.003.0009

Audio can have a special kind of magic that stands apart from print and film. It's more intimate. A good story—especially one that features a compelling voice—can help us see each other more clearly and challenge what we think we know. Audio is vivid. Even a small detail can transport us somewhere else. It can be more accessible, too. A well-done podcast can turn a complex process, like the development of a coronavirus vaccine, into a riveting journey full of unexpected lessons, featuring the scientists at the center.

And science—with all its questions, mysteries, and explorations—is a perfect match for this medium.

The Basics

What exactly *is* a podcast, anyway?

The world of audio journalism is incredibly dynamic. Just click on the health and science sections in iTunes to see for yourself. From daily science podcasts like NPR's *Shortwave* to season-based shows like Gimlet Media's *Science Vs.*, the list is long. It's growing. It's evolving. Who knows what might have hit the scene between the time I write this essay and the time that you are reading it?

Storytelling may be as old as humankind, but podcasting is a young medium. It was only in 2014, seven years ago, that *Serial* from *This American Life* came out and shook up the industry. The series garnered millions of listens, unheard of before, and proved to risk-averse public radio leaders and energized producers that an appetite existed for longer form storytelling and audio programs independent of radio.

It helped create new spinoffs, startups, and a golden era for the field. Entire podcast production companies have emerged, and limited-run series are all the more common, like WBUR's *Infectious* and NHPR's *Patient Zero*.

Print and online publications have also started adapting podcasts, as have nonjournalism groups like medical associations and educational institutions. The health and science beats, which involve producing a mix of short- and longer term features, remain a central part of more traditional radio news outlets like the BBC, NPR, and local member stations.

As a result, an audio reporting project can take many different shapes and forms—and increasingly, may be part of a multimedia package that also includes an online story and visual element. The range and scope depend on the show, your own vision, and the timeline.

Producing a segment for a live, interview-based program like *Science Friday* may involve identifying guests and outlining questions for the host. Longer interview-based shows and podcasts, like *Fresh Air*, are prerecorded and require cutting down—the common term for editing audio—your interview.

Contrast that with the kind of 4-minute story that typically appears on a program like *All Things Considered*, with narration woven in and out of excerpts of interviews with researchers and patients. Or contrast with a deep dive into the history of a drug from its indigenous beginnings to its present-day uses, often drawing surprising new connections, for a show like *Radiolab*.

Whatever the format, it's essential to know what kind of story you want to make and who your audience is. This will help inform everything from how you organize a story and approach the reporting, to the questions you ask sources and the depth of those conversations.

Getting Started

Podcasting and audio storytelling have a serious DIY element. All you really need to get going is a recording device of some kind, computer software to edit the audio (some programs are even free), and, of course, curiosity about an idea or a question you wish to pursue.

Resources and how-to's abound. The Association of Independents in Radio (AIR) has guides for pitching stories, suggested pay rates, and mentorship programs. Transom.org contains myriad essential recording hacks, essays, and even a podcast, *HowSound*, that deconstructs how some of the most moving audio is made.

NPR training guides are full of gems about putting stories together. The Third Coast International Audio Festival houses a complete treasure of archived lectures and curated mixes that feature some of the best work out there, with makers generously sharing their secrets of the craft.

Radio stations and podcasts are increasingly offering paid internships and fellowships. Additionally, radio clubs and listservs are becoming more and more popular, depending on the city or region where you live. This can be a great way to connect with other makers, to learn and to conspire. If a group doesn't exist, try setting one up. Organize a listening or feedback session on a specific theme, on works in progress, on pieces that inspire you—or if you need some creative inspiration, take part in an audio playground assignment (a monthly assignment from Chicago-based audio maker Sarah Geis) and share the results.

Exploiting the Medium

With each piece I'm planning, I'm constantly thinking about what makes this story worth *listening to* as opposed to reading or watching. What specific elements can really make a story sing, so to speak, through radio? Would a patient be willing to record an audio diary? How can I capture the emotion involved and identify someone who may be willing to recount an event or describe an experience? Is there an interaction between people I can document? Is archived audio or online material available to help move the story forward and add more depth?

One of the best parts of the audio reporting and production process is finding good "tape," a term that harkens back to when audio was recorded on reel-to-reel tape machines. It's still the common phrase for audio, even if it has long turned digital. Good tape is the life source of an audio journalist, but beware, it doesn't necessarily mean a literal sound, like that of a scientist flipping the pages of a research paper or the sound of a school bell (cliché alert!).

It can be the raw way that a source recounts an experience; the authentic, impromptu phone call a doctor receives during the interview that demonstrates his or her work; or the way a person describes the photo of a loved one. Finding these gems is an active process. It requires envisioning what you might want, how it might fit in your rough story outline, and then seeking it out. Along the way you may wind up with something unexpected and better.

After returning from the field or an interview, you might wonder how to decide which parts to use. I'm always taking notes, even if it's just mentally, during the interview itself. Right after you get back, it's useful to write down or tell a colleague about the moments that stand out. This can shape the structure of the story, strengthen your preconceived outline, and sharpen your ear when going back through the recording.

You very well might wind up with too much good stuff. A lot of what you record won't make it into a story or podcast. It's painful. I still think about parts of that conversation with Portugal's drug czar that never made it in. But in the writing, editing, and production process you'll gain a clearer sense of what needs to go. Having a story outline ahead of time really helps in making those tough decisions. How does a great moment of tape reinforce and fit in the broader story? It doesn't? It'll have to go. Does it repeat a point that was already included? Leave it on the cutting board.

There are potential pitfalls along the way. Audio storytelling is linear. It can capture one's attention through a chronology of events. What happened next? And then what? The inclusion of a small detail can make a piece memorable for years. But too many details, plot lines, voices, and information can confuse the story and risk losing listeners' attention. Be disciplined in what is necessary to move the story forward and what can go.

It's Both an Art and a Science

When I worked on WHYY's *The Pulse*, a weekly health and science show, we were always looking for ways to pull back the curtain on research and healthcare. We wanted to identify stories that really drew listeners in. For the team, that meant gaining access to a lab or clinic. Shadowing and recording scientists in addition to sit-down interviews, finding important voices—those impacted or at the center—who were willing to be interviewed, who'd let us follow their journey.

Invisibilia cohost Yowei Shaw stresses that many factors play into story decisions. Whether a pitch gets the green light can be determined by a lot of things. She lists: "Your background and positionality to the story, the conventions of the industry of the time, the conventions and goals of the show at the time, available resources, the vibe in the pitch meeting."

Traditionally, we're taught that a powerful narrative radio story has elements of surprise, it has stakes, it has compelling characters who have something interesting to say about their experience, and, depending on the show, it has a good news hook. Yet a story can work well even without a straightforward plot. Beautiful writing and production, a powerful question, an unusual tempo, and dramatic enough stakes and buildup can make all the difference.

"I really do think you can make anything work," Shaw says, "if you could come up with an inventive way to do it in the right format."

Pay attention to the formulas and tools of audio storytelling, but leave room to push boundaries and try different approaches. For health and science, specifically, audio provides a chance to humanize the research and the complexity. That's because audio stories are inherently conversational and can help bring numbers and data to life.

On Interviewing

Certain interviewing techniques can be especially helpful for radio and podcasts. I'm always thinking about the listener being present and how to keep them engaged. The last thing I want is for someone to get lost or caught up in complex explanations, unfamiliar jargon, or too many details and numbers. That runs the risk of listeners tuning out or, worse yet, turning off the program.

That's why it's important to set a comfortable, conversational tone and to be willing to ask simple questions. Not only be curious and follow up on the responses you get, but also be ready to ask a question again if the answer wasn't clear or if it seemed stiff. Try paraphrasing a response and repeating it back to an interviewee to see if that might conjure a sharper description or an unexpected reaction.

Finally, being conversational and clear is just as important in the scripting of a story or any narration as it is in the interview. Write in short, clear sentences and read the copy out loud before an edit or recording. Does it seem like something you'd say, or is it awkward? You'll know if you stumble.

Working With Sound

You might have the best story in the world, and even the best access and interviews. But guess what? None of that matters if the recording that you did doesn't sound good. It may as well not have happened. Imagine someone sharing the most intimate detail of a research trip, only to have an ambulance outside drive by, blaring its alarm at a very exciting moment in this story. (It doesn't hurt to ask your subject to repeat what they just said.)

If you can't meet a source in person, figure out the best remote recording setup for capturing the clearest sound, whether that's arranging for your interviewee to go to a recording studio; hiring someone to record the interviewee's end of the conversation; having the interviewee record on a smartphone or computer and then sending you the file; or at the very least, having the person you're speaking with find a quiet place that has a clear phone connection (and be sure to check that the phone is not on speaker!).

When recording in person, different microphones have various settings that may require handling them in different ways, in many cases within inches of a person's mouth. It's important to get used to doing this and ensure that a person is "on mic." The more comfortable you are, the more comfortable the person you're interviewing will be. It's amazing how quickly that microphone disappears when the focus shifts to the interview itself.

It also might seem tempting to take off your headphones, but wear them at all times when you're recording. That's because they will tell you what your recording actually sounds like later. Otherwise you might be inside a lab and not notice the loud, distracting buzzing of a machine and fail to move the interview to another, quieter side of the room.

Sometimes you can't escape sounds, and you might be outside or next to that buzzing machine for good reason. An essential editing trick is to always—I repeat, *always*—record several minutes of what's called "ambi" wherever you've recorded something. Ambi refers to the natural sound of a space, of chatter in a room, of a river flowing, or of birds chirping. Even if it doesn't seem loud or sound like anything is happening, every space is different. Recording ambi will be extremely helpful in postproduction to build scenes and smooth out transitions in and out of audio clips that were recorded in that place.

A microphone can also act like a camera lens, with close ups, middle shots, and panoramas. If there's a rehearsal, try getting close to a particular musician, and then step back to record the group as a whole.

In the case of my meeting with Portugal's drug czar, the former drug market was right next to a highway. Not an ideal place for an in-depth interview! But I knew that morning was the only chance I'd have to meet with him and I wanted that scene with him outside. So after I recorded our walk and captured several minutes of ambi along the way, we looked for a quieter spot to continue talking. We couldn't find one (the nearby cafe was way too noisy), so we wound up doing the interview in his car, which was parked in a quiet parking lot. The lesson? Plan ahead as much as possible, but be ready to improvise.

The Takeaway

There are so many audio possibilities for health and science journalism. The field is extremely accessible if you're willing to learn some basic recording techniques. Audio is also a highly collaborative medium. Like science itself, podcasts and radio projects often involve a team of producers, reporters, hosts, and editors, who together make a story even better. Within this rich ecosystem is an opportunity for you to find your voice. So start experimenting.

9
Film and Video Storytelling

Ian Cheney

Ian Cheney is a Peabody Award–winning and Emmy-nominated documentary filmmaker. His films include King Corn, The Greening of Southie, Truck Farm, The City Dark, Bluespace, The Search for General Tso, The Most Unknown, The Emoji Story, Thirteen Ways, Picture a Scientist, *and* The Long Coast. *He has also created short films and series, including* Moon Mirrors *and* The Measure of a Fog, *and his films have appeared on outlets including Netflix, PBS, the Sundance Channel, and NOVA. A former MacDowell Fellow and Knight Science Journalism Fellow at Massachusetts Institute of Technology, he received bachelor's and master's degrees from Yale University and an MFA in film from the Vermont College of Fine Arts. He lives in Maine.*

In my darker hours, I've grumbled to myself that science filmmaking is impossible. Science brims with complexity and details; filmmaking demands a certain narrative simplicity. But invariably the fog lifts, and I'm reminded why this unusual intersection draws me in time and again: Science is a journey, and journeys make for good storytelling.

I mention this at the top because there's a certain pressure in science storytelling to recite results, explain theorems, and list discoveries. Fair enough—there's a lot in the history of science to be proud of and share, and lots of science reporting hinges on relating the latest news: Here's what happened, here's why it's important, and so on. But if we keep in mind that science is also a process, propelled by human beings and powered by curiosity, then suddenly the explainers share the stage with an equally enthralling—and revealing—narrative. Scientists are professional wonderers in a universe full of unknowns, and filmmaking gives us a chance to go along for the ride.

Documentary audiences demand vibrant, emotional storytelling because of the medium's deep roots in fiction cinema. Luckily, science abounds with compelling narratives, wondrous visuals, and fascinating characters. The foundations of good storytelling are there. And both science and filmmaking enhance our ability to see the world in new ways. A simple image of a telescope

Ian Cheney, *Film and Video Storytelling* In: *A Tactical Guide to Science Journalism*. Edited by: Deborah Blum, Ashley Smart, and Tom Zeller Jr., Oxford University Press. © Oxford University Press 2022. DOI: 10.1093/oso/9780197551509.003.0010 can evoke the realm of distant galaxies; footage of an astronomer alone in a mountaintop observatory illuminates the deeply human quest to understand the stars.

For scientists, the spirit of innovation might be second nature, but for video and film storytellers working on a deadline, it can be hard to keep in sight, and you can end up with a lot of shots of scientists awkwardly walking down hallways in their lab coats. Fresh approaches to science filmmaking are pearls in an ocean of mediocre videography. It's worthwhile to watch other films to get ideas for what works and what doesn't, but the best way to learn is to start making your own mistakes. Film a little, edit a little, watch it through with colleagues, and repeat. Doing this when you're not on deadline is an excellent way to experiment with new approaches, like filming with a handheld camera or using special equipment like a dolly or a jib. Eventually you might find yourself hiring a crew of specialists, but the more intimately you personally understand the different moving parts of filmmaking, the better.

In other words, filmmaking, like science, is both a journey and a process. Keeping that in mind helps plant you firmly in the storytelling mindset: beginning, middle, and end.

Preproduction

The first interview I ever filmed was with an Amish man in Ohio. A beloved nature writer and hard-working farmer, he had responded to my colleague's interview request with a handwritten letter inviting us to his small homestead. But when we arrived, it was clear that I hadn't done my homework. He was happy to be filmed, but for religious reasons we would not be allowed to film his face.

"Film my hands," he suggested, and we agreed this seemed like a sound plan: His hands were weathered and worn, a perfect metaphor for all he'd seen in his life. As we set up the camera, I recall thinking my filmmaking career was off to a bold and innovative start; I'd never seen an interview of someone's hands! We sat him on a wicker chair in the dappled shade beside his barn, he placed his hands comfortably on his thighs, and I filmed an hour-long interview of a farmer's crotch. It didn't make the final cut.

Thinking ahead is obvious, but it's nonetheless hard, largely because filmmaking is multidimensional. You have to simultaneously research your characters, topics, storylines, camera gear, audio equipment, weather, location, timing, travel—not to mention the dietary restrictions of your crew. It can feel like a crazy constellation of things to keep track of, especially when you're gazing into complex scientific terrain.

So take a step back. Everything in filmmaking revolves around a linear timeline. What happens at the beginning of your piece? The middle? The end? Sketching out scene ideas and storyboarding your overall sequence of scenes will help you shape what you need to capture in the field. I often enlist an informal board of advisers at this stage. They serve a few purposes: keeping you on track scientifically; introducing you to key contacts; challenging your assumptions; and pushing you in new directions narratively. I approach a few people from the sciences, a few people from the arts, don't bother them too often, and offer them a credit in the film.

The single most important part of preproduction is finding your film's characters and working with them ahead of time whenever possible. We don't tend to use the word "casting" in nonfiction filmmaking since we're not working with paid actors, but it's essentially the same idea: Who is going to be in this movie? (And can you film their face, or just their hands?)

Scientists have been using video calls for years and are accustomed to bouncing around ideas. Introducing a certain spirit of collaboration from the get-go—we're making this thing *together*—tends to reveal new possibilities and ward off potential snags. Plus, forging a personal connection will go a long way toward building the trust necessary to make filming work.

A sense of what your film will look and feel like also comes into focus during this early phase. What's the right tone, the mood? Will you film handheld, with quick cuts? Will there be both sweeping drone shots and intimate close-ups? In addition to the footage you shoot, what other visual elements might you include, like archival footage or animation? Grab images from the internet to make a sample collage of what your film's style might be and use that as a guide. Think about how to avoid stale or overused techniques from past science films and try to come up with a visual approach that is tailored to your topic.

And finally, it's time to pull together a crew. If you haven't made many films, you might assume you can do everything yourself, or conversely you might assume you can't do anything at all. Neither, of course, is the case. The first step is to clarify your strengths and interests: What makes sense for you to do, and what jobs do you need help with? It sounds obvious, but hire a crew that complements your talents and fits your film's needs. If the film has challenging logistics, bring aboard a producer to help wrangle details. If the film involves challenging physical locations, make sure you have a production assistant to help carry the tripod. Also, skip the fancy lenses and hire a sound recordist. They're worth every penny.

Shooting Your Film

Some time ago, a few colleagues and I were setting up to interview an astronomer at a major observatory in the southwestern United States. Our cinematographer noticed that there were a lot of dusty old PCs making annoying sounds in the room where we planned to film, and he set about unplugging them since they didn't seem to be connected to any monitors or otherwise show any signs of regular use. The interview we filmed was wonderful, but the PCs were in fact the hubs of astronomical data connecting the big telescopes on the mountain to astrophysicists around the world. Apparently we didn't cause any damage, but the lesson was clear: Next time try not to destroy science while making a movie about science.

That's not a bad place to start when thinking about how to shoot your film: In what ways can your production not only not harm science, but also possibly benefit the project of science? It isn't easy making scientists comfortable working with a film crew since the experience can be annoying: Walk down this hallway; now do it again. Take off your shirt for a second and put this microphone on. Would you mind repeating that five different ways, and is it OK if we rearrange your laboratory to make it look better for the film? Can you summarize your life work and the history of the universe in two snappy sentences?

The best way to avoid these annoyances is to let the scientists in on your process before the shoot. Tell them how you're thinking of constructing your story, how your equipment works, what your challenges are. There are more similarities between storytelling and science than you might think, and perhaps the process can spark unexpected opportunities.

I made a film a few years ago called *The Most Unknown*, where I invited a scientist from one discipline (say, microbiology) to visit the laboratory of a scientist from a very different discipline (say, astrophysics) and filmed what happened. It wasn't intended to be an opportunity for direct research collaboration, and it did not become such, but several of the nine scientists reported back to me that the experience was deeply positive, allowing them to immerse themselves in another discipline and see science more broadly and in a fresh light. As a filmmaker, it showed me the power of experimentation, and it led to several new projects, including one where I invited 12 people, mostly scientists, to walk a plot of land they had never seen before and describe what they saw. How often does a scientist get to step out of the lab and take the time to observe?

Be patient with the people you're filming and interviewing and be open to a certain amount of improvisation. Since much of contemporary science is deeply specialized, you might not understand everything a scientist is saying, but let them say it all the same. The best stuff is often the stuff *they* are excited about. Showing them that you're there to really listen will open doors and deepen your collaboration. Filmmaker Jason Rosenfield told me that's the best advice he's gotten about directing a documentary: *Be the best listener your interviewee has ever encountered.* Once you've given someone a chance to describe the world as they see it, it's easier to go back and have them explain things a few different ways, giving you options in the edit.

You don't want to be saddled with more footage than you can possibly process, but it's hard to go back later to get footage you missed, so err on the side of filming just a little bit more than you think you need, and you'll be glad for it later.

That said, I've come to avoid the term "B-roll," the customary slang for the footage that complements and supports the main footage. That's the wrong way to think about it. There shouldn't be any shots in your film that aren't awesome. If you need an exterior shot of a laboratory building to establish the setting, put some thought into how you frame the shot, or what time of day it is, or what might be going on in the foreground. Science teems with wondrous sights if you're willing to take the time to look. Use a tripod, get good sound (without unplugging the scientists' computers), and bring the world to life.

Postproduction

Perhaps it was cosmic retribution, but a year after unplugging the PCs during my shoot at the observatory, I lost an entire interview I filmed with an astronomer in Hawai'i. I still have no idea how it happened—I'm maniacal about backing up my footage. Fortunately, this doesn't happen very often since when it's gone it's gone, and we weren't able to reshoot the interview.

Any scientist will tell you that their data are precious. Footage is our data, so back it up. Label it clearly. At the end of a day of shooting, write a log of what happened during the day, corresponding to the footage organized on your hard drives, and include any notes about special moments or technical glitches. This can be essential in the editing room months down the line.

Once I've successfully carried my hard drives home and stored copies in different locations, the next step is to make transcripts. Given the complexity of a lot of scientific topics, taking the time to make transcripts of your interviews and conversations will vastly enhance your ability to lay out a coherent story.

The next step after getting home is to hire someone to look at your footage with fresh eyes. An editor won't share your same biases and will only see what's in the footage, not what you think is there because you've been planning the film for so long. A good editor will help you realize your vision while also elevating it to a new level. Like science, filmmaking is a team sport, even if there are strong personalities involved.

As you wade into the edit, you're really wading into the storytelling, translating everything you've done into a linear timeline. It's at this stage in particular that you have to keep your audience in mind. What do they know? What are they wondering about? Many fiction films have complex plot lines, overlapping narratives, and small details that get pieced together later; science documentaries can be the same. Trusting your viewers' intelligence doesn't mean using incomprehensible jargon or piling on facts at breakneck speed: Clarity is still essential. But the more you invite viewers to synthesize an understanding of their own, the deeper their engagement with the science will be.

If graphics and music deepen this engagement, terrific. But avoid leaning on these elements to prop up a lackluster story. Lousy science content often gets a layer of relentless music strung under it or flashy whiz-bang graphics laid over it. These are empty calories. Take your time here and make sure everything clicks together as a unified piece.

Taking your time is a point worth repeating. Editing is the lengthiest and most difficult part of the documentary filmmaking process. Rushing rarely helps. And for science storytelling, there is often a lot to be discovered in the footage: little moments that add human subtlety to a team's experience in the field; side remarks that shed light on what a scientist is experiencing during research; or a key visual that helps clarify a part of the scientific process.

The Takeaway

Film is best suited for conveying stories and inviting further thought—not as a vessel for transmitting encyclopedic amounts of information. If scientists are your main characters, let their humanity shine through; don't get bogged down in trying to regurgitate every detail of their research.

Try to be as original and innovative with your storytelling as the best scientists are when designing a new research study: build on what came before you, but look for a chance to break new ground. And above all, find joy in the work. Science offers opportunities to see your world anew, and if you embrace that spirit of discovery it will shine through every frame of your film.

10

Multimedia Storytelling

Jeffery DelViscio

Jeffery DelViscio is the head of video and podcasts at Scientific American. *Previously, he was a Knight Fellow at the Massachusetts Institute of Technology and the founding director of multimedia and creative at STAT, where he oversaw all interactive journalism, video, audio, photography, and social media. He spent over 8 years at the* New York Times, *where he created and edited multimedia across five different desks, working with correspondents around the world. Before entering journalism, he worked aboard oceanographic research vessels and tracked the influence of money and politics in science from Washington, D.C. He holds dual master's degrees from Columbia University in journalism and in earth and environmental sciences.*

In 1961, Morton L. Helig unveiled the "Sensorama Simulator," an invention that combined "the cooperative effects of the breeze, the odor, the visual images and binaural sound that stimulate a desired sensation in the senses of an observer." It never quite took off, but the seed of the idea did. We can see its DNA in today's smartphones and virtual reality headsets. (The "odor" enhancement has been, thankfully, taken off today's features lists.) Helig's Sensorama arguably marked the beginning of the digital multimedia era.

To journalism, multimedia is the promise of a unified, seamless storytelling experience that hits more than one of your senses at a time. It's about breaking out of static web pages and expectations. It's about surprise and content that glues eyes to the screen. And it's about the targeted use of the right storytelling medium for the right narrative moment.

The news industry has traveled a rocky road to embrace multimedia over the last decade. There will always be challenges to getting this kind of storytelling done in a newsroom, but to take up that challenge is to help journalism evolve in its ability to meet the changing news needs of future generations across our increasingly fractured information landscape.

Jeffery DelViscio, *Multimedia Storytelling* In: *A Tactical Guide to Science Journalism.* Edited by: Deborah Blum, Ashley Smart, and Tom Zeller Jr., Oxford University Press. © Oxford University Press 2022. DOI: 10.1093/oso/9780197551509.003.0011

A Short History of Disruption

Journalism didn't see or appreciate multimedia's possibilities until the last decade or so. Our slow adoption was part structural, part dogmatic.

Newspapers started in colonial times, radio in the early 1920s, and broadcast in the 1940s. For much of the last century, these disparate media grew as specialized formats with little overlap. Journalism schools reinforced these silos in their curricula to help their students get jobs in each specialization.

Around the time I attended journalism school in 2004, something had started to change. A new category, strange and amorphous, appeared: "new media." The initial embrace of multimedia had begun, spurred on by the explosive growth of the internet. But another new force, riding the back of that surge, would be the gut punch that journalism didn't see coming: the rise of the platforms.

Facebook launched in 2004, YouTube in 2005, Twitter in 2006, and Instagram in 2010. These platforms represented a radical transition in the way people consumed news and shared information. Journalism's historic power as a solitary gatekeeper of information was lost, almost overnight.

Platforms sucked in users (along with their personal data) by the millions and began to leverage those data to steal attention and advertising dollars from the media. As we covered the phenomena of social media, journalists largely formed a defensive crouch against the platforms' corrosive onslaught. Waves of cuts, layoffs, and downsizing would wash over newsrooms in the years that followed.

At the same time, the internet became able to transmit audio and video. Diminished newsrooms had to pick up unfamiliar technologies and ways of communicating. It was the beginning of an ever-pivoting race to keep up with a younger news consumer, whose expectations about content had completely changed.

Inside newsrooms, something organic was happening. Younger staff brought new skill sets; they built more flexible content management systems and filled them with audio, video, and interactivity. These changes kicked off a new era of storytelling experimentation that would come to define what journalism has evolved into today.

If you're someone who's just going into our business, you've arrived after the first waves of disruption—waves that broke into a new, though still unsettled, normal. Multimedia isn't a foreign concept. You move fluidly between text, audio, video, and interactivity on many digital platforms. Your generation had digital flexibility from the start.

So, You Want to Make Multimedia . . .

None of this means it's easy to get multimedia done in the newsroom. You will likely land at an outlet that's understaffed. Thinking in multimedia requires even *more* work than traditional reporting. And if you're freelancing, it's even harder to convince an editor you can deliver multiple media *well*.

It is helpful to fixate on two primary challenges: your storytelling dexterity and the opportunities (and limitations) of your outlet.

Let's start with the one you can control—you.

When I got to the *New York Times* in 2006, I was almost immediately labeled the resident "audio expert." Why? I had taken one class in radio reporting in graduate school and had, only months earlier, learned to record and edit audio. My expertise was shallow, but I showed no hesitation in leaning into the title. My colleagues didn't know how little I knew, and I made it my job to mature that knowledge quickly so that they never would and the moniker would stick.

Get used to learning new skills, and quickly expand the ones you *sort of* have. Don't know how to work a digital recorder? You will in a few days of fumbling about. Never held a video camera, much less shot with one? Yes, you have—your phone. Treat it like a proper camera. Stabilize it with a minitripod. Add a low-cost, plug-in wired microphone for interviews. Learn all of its manual modes to control exposure and maximize video quality. Your phone can help you learn how to shoot and edit video proficiently.

Quickly develop *taste* in multimedia—identify what's good and, even more importantly, what's not. Spend too much time watching animation technique tutorials on YouTube. Sign up for newsletters from creative media makers of all kinds; they offer valuable training resources and surface others' good work.

Try to become obsessed about learning one specific thing: how to light a subject properly, how to fly a drone, or how to log tape efficiently. And then set yourself up to do it over and over until you have mental muscle memory. Then move on to the next obsession. (If you want to obsess about *areas* of multimedia production, rather than just specific *skills*, the video, audio, and data storytelling chapters in this book are great places to start.)

The other challenge is more out of your control: your newsroom and its mentality toward multimedia.

Your outlet will likely opt to do what's cheapest, fastest, and most trackable (via some key website performance metric). Part of getting multimedia *done* is understanding this pattern—and the opportunity it presents. Overworked editors may welcome multimedia experimentation if the work shows results in clicks, story dwell time, and journalism awards.

You may find yourself collecting multimedia while reporting. Conduct your interviews with an eye and ear open to how they could look or sound in your story. Recordings can quickly be turned into embedded audio, especially if they are high quality. You would be surprised how useful gifs can be for showing wondrous or process-oriented things. (Create these by looping together a phone camera picture burst or time lapse of some discrete, story-related visual.)

But before you ever push a red record button or depress a camera shutter, you need to answer some key questions:

Does multimedia even make sense for this story? Multimedia can excel at helping readers form a better mental picture of a complicated story. When leveraged correctly, it augments the experience. But multimedia should work in harmony with the text—not be a narrative duplicate in a different skin. Only experience will help you understand how to achieve that balance, so don't be afraid to experiment (and fail at it sometimes).

Once I was tasked with coordinating a short documentary *and* a long-form article on a gripping opioid death where a man gave his best friend the dose that killed him. I pushed my team to ditch the documentary, so that video could bring out visual details within the text. The result was a 3,500-word story with 21 short video loops. They were placed contextually to seamlessly weave the most evocative visual moments into the narrative flow. The piece was viewed by 250,000 readers who stayed with it for 15 minutes on average—a level of engagement miles above the site's typical pattern.

Where does it fit in the story process? Optimally, multimedia should be being considered at the outset of a project, rather than as an add-on later. Starting early will help you to properly decide on the best mix of text, audio, video, or interactive media to maximize story flow. This will also help you figure out if you can collect all the media yourself or if you will need to involve others in the newsroom.

Multimedia may also require field reporting and additional technical skills. Photographs can often be made or secured relatively quickly, but even if you can quickly collect video or audio, the postproduction process can be complicated. You need to build in time to let these processes play out. And if you start late, you will likely have to give up on more technical and time-consuming additions, like reported video or animation.

What will the final presentation of the story look like? It's gotten a lot easier to publish multimedia in the last decade. Everything used to be built using custom-made computer code. Now, many organizations have developed their own display templates. This can speed multimedia production significantly.

Find out if your newsroom has these tools and how to use them to build any project you're thinking about. Get familiar with their bugs. Befriend your colleagues who built them or use them often. You'll need help and guidance. And if you need some functionality that doesn't exist in-house, chances are you can find a tool that does what you want on the web, complete with flexible embedding code.

More than anything, don't let your first project be the first time you pick up your newsroom's multimedia publishing tools. It's hard enough to write a story on deadline, let alone learn new software.

What's your target audience, and what will the story rollout look like? When you get to publishing time, think about which pieces of media will communicate your story most clearly to an endlessly scrolling audience. Having several formats in hand will let you appeal to different audiences on different social media platforms—more bites at the algorithmically modified social media apple.

Science Meets Multimedia

This is a tactical guide to *science* journalism, but so far this chapter has focused on general multimedia concerns. There are, however, some important specifics when it comes to science:

Research has a data trail. Science generates a lot of media (photo/audio/video), sometimes as part of the actual research, sometimes as a chronicling of the process. This could be anything from GoPro footage to ambient field sound to visual simulations to slow-motion video. These can be compelling additions, used appropriately and contextually. You will not know that they exist if you don't ask your sources what they record and why.

Beware of the deer-in-headlights researcher. Many scientists will be unprepared to sit in front of a camera or speak into a microphone. Prepare yourself for them to be awkward and unsure. Have several conversations, if you can, before seeing them in person or visit several times to make the assessment before filming or recording. The extra time will help put them at more ease—or convince you that perhaps this source isn't the best choice for video or audio. Academic press offices are filming their own researchers more often now. Perform a preinterview Google video search on your source—it's a good preparation tactic, even if it turns you off them for multimedia.

Beware science's "nothing to see here" problem. Many labs end up looking exactly the same—same lab benches, same nondescript machines humming away doing something both wildly complicated and shielded from view. Unless you are an animator, or have access to animators, videos shot in labs can be visually boring and fail to use the medium for its best purpose: to show, not tell.

Conversely, lots of science is truly invisible but has compelling visual possibilities. How does one tell the story of a pulsar or a tiny round-worm being used to understand the roots of cognition? You could de-scribe the rotating star, emitting beams of electromagnetic radiation out of its poles. Or you could listen to the heart-like thump, thump, thump of pulsar B1933 + 16. You could write about how *Caenorhabditis elegans* has only 302 neurons with 7,000 synapses—presenting a vastly simplified anatomical system that helps us learn about our own wild complexity. Or you could show a three-dimensional flythrough of that worm, pausing in virtual flight to pick out individual neural ganglia and show why they're important.

Remember how I advised obsessing over some multimedia skill? If it's only one, spend 6 months getting to know Adobe's After Effects. It's an animation and compositing program, and it's hugely powerful. Learned well, it can help you bust the lack of visuals problem wide open.

The Future of Storytelling

Can you imagine what a news story will look like 20 years into your career? Will the small and large screens that dominate our lives now still even exist? Will social media evolve into something even more invasively connected to our every thought and decision—perhaps literally, flowing through neural implants?

Now look two decades back. There were no iPhones, the word "selfie" didn't exist, no Google, no Twitter, no Facebook. Netflix was a DVDs-by-mail company. Streaming? Not a thing. Your news came mostly on flattened, dead trees or maybe you watched the nightly news . . . well, nightly.

That has all changed. Social networks connect us—and drive us apart. Misinformation spreads like digital wildfire. Journalism seems forever on the ropes as the platforms aggregate our content and monetize every click for their own gain.

Where does all this leave multimedia? In demand. Many journalism jobs today require it. A 2019 study of 669 full-time job listings for journalists showed that 85 percent of the listings required multimedia (a very close second to writing).

In a larger sense, multimedia is one of the remaining hedges journalism has against obsolescence 20 years from now. To learn how to wield multimedia is to divine the path of where journalism needs to go.

What will the users of 2040 look for in their news? Will they retreat further from the mainstream media into niche digital spaces that reinforce their own biases? Or will you be prepared to use any storytelling means—text, video, virtual reality, some format yet to be developed—to reach at least some of them before they do? Will you innovate and evolve what it means to be a journalist, no matter the medium?

The Takeaway

Multimedia has become a critical component of science journalism's storytelling arsenal. When leveraged correctly, it can increase reader engagement and social transmissibility of a story. Science is replete with opportunities and source material for making multimedia.

Training yourself to create it may seem daunting, but, if you are fearless and flexible, you may quickly become the multimedia expert in your own newsroom. In this age of real resource limits, that expertise will be important for getting ambitious storytelling done.

11

Data Storytelling

Charles Seife

Charles Seife, a professor of journalism at New York University's (NYU's) Arthur L. Carter Journalism Institute, has been writing about physics and mathematics for almost three decades. He is the author of seven books, in- cluding Zero: The Biography of a Dangerous Idea *(2000), and, most recently,* Hawking Hawking: The Selling of a Scientific Celebrity *(2021). Before arriving at NYU, Seife was a writer for* Science *magazine and had been a U.S. correspondent for* New Scientist. *His writing has also appeared in* The Economist, Scientific American, ProPublica, The Philadelphia Inquirer, Discover, Slate, Smithsonian, The Washington Post, *and* The New York Times, *among other publications. He has also been a scientific consultant and writer for television documentaries about science and mathematics. He lives in New York with his wife, Meridith, and his children, Eliza and Daniel.*

Data sets are sources that don't mind if you call them up at 2 in the morning. Ask them the right questions and in the proper manner, and they'll answer without evasion, deception, or ulterior motive. They can help you find stories that nobody else has—as long as you can track them down and use them properly.

Just as there's a learning curve to cultivating and interviewing human sources, it takes a bit of time and practice to figure out where to get data that bears on a story and to get good at cleaning and interrogating data sets so that they reveal useful information. But, it's well worth the effort.

Where to Find Data

There's an incalculable amount of data at our fingertips, sitting there, waiting for us to download and use. We just have to know where to look.

Nobody hoovers up more data than governments. Wherever there's a regulatory body, wherever there are people watching and governing the behavior

Charles Seife, *Data Storytelling* In: *A Tactical Guide to Science Journalism.* Edited by: Deborah Blum, Ashley Smart, and Tom Zeller Jr., Oxford University Press. © Oxford University Press 2022. DOI: 10.1093/oso/9780197551509.003.0012 of businesses and people, there's vast amounts of data. And now, many of those governments make that data easy to get.

Whenever there's a federal agency relevant to your reporting, odds are they have a database that will be of interest. In the United States alone, the Environmental Protection Agency tracks pollution and air and water quality; the Food and Drug Administration gathers data on clinical trials (along with National Institutes of Health [NIH]), drug and medical device adverse events, and food ingredients; and the Nuclear Regulatory Commission keeps an eye on nuclear power plant safety. The Department of Agriculture compiles data on agricultural productivity; Fish and Wildlife has a database of threatened and endangered species; NASA has climate data; the U.S. Geological Survey has data on earthquakes; and the Centers for Disease Control and Prevention's (CDC's) Wonder database records how people die.

That's just for starters. Data are for not only the feds, but also regional and local governments. Many forward-thinking smaller governments have taken to creating central clearance houses for all of their data sets; New York City's Open Data repository, for example, contains a bewildering amount of data regarding its residents and the environment they live in.

And if government data happen not to be online, you can often request them; the federal Freedom of Information Act and its many regional analogues typically entitle you to many of the data that these governments collect, even data that they're not keen to share.

Government data are just the beginning, though; the entire internet is made of data after all. Sometimes, you'll find data sitting on a website, all ready to download directly into your spreadsheet.

With a little more skill, you can write a computer program to pull data off of websites and make it available for your use. Many data journalists have mastered this skill—"data scraping"—to generate interesting data sets. If you're lucky, though, you don't have to go through the trouble of grabbing and formatting data from a website. Many data sets stored on computers can be accessed via what's known as an application programming interface (API). I like to think of an API like a data faucet: hook up to it in the right manner, turn the handle, and data pour directly from the server into your computer like water.

Search engines have little-used features that can help you find interesting data sets. Google allows you to limit your search to certain kinds of files (e.g., Microsoft Excel spreadsheets); combined with a domain restriction (e.g., to government websites), this turns the search engine into a powerful data-finding engine. For example, typing "oil spills" filetype:xls site:gov into a Google search box instantly yields a variety of relevant data sets from the Bureau of Transportation statistics, NASA, and other federal and local agencies.

It's also possible to create your own data sets; anything that you can observe or extract from a source and record in a spreadsheet or database is a potential source of data for a story. A few years ago, *Slate* used media reports to create their own data set of incidents of gun violence in the United States. As imperfect as it was, the database became a powerful instrument for gathering information about a public health problem the CDC was unwilling to study.

What to Do With Data

Most data sets are dirty. And all but the most carefully curated databases are worse than that—filthy with bad data, poor formatting, and inconsistent recordkeeping. So a likely first step is to roll your sleeves up and dig in: root around in the data to understand its quirks and get a sense of what is reliable and what isn't. If you're lucky, the errors will be small enough that you don't have to worry so much about them. If you're not, you might have to spend some time cleaning up—correcting or excluding erroneous data. In any case, you'll have to ensure that the conclusions that you want to draw are robust enough to survive the messiness of the data.

For example, every so often, you'll come across a data set containing a wildly implausible number of centenarians. On closer inspection, the vast majority of the centenarians usually have a birthdate of January 1, 1901. This isn't a sign of extreme longevity. It's a quirk: Some databases try to handle missing birthdays by filling them in with a default value at the beginning of the 20th century. So every a time a data entry clerk entered a record with a blank birthdate, the database turned it into a new individual born in 1901.

Sometimes compiling, cleaning up, and organizing a data set—putting it in a nice, neat, easily accessible format—is enough to be the engine of a news story or several news stories. A few years ago, *ProPublica* collected publicly available data about pharmaceutical industry payments to physicians. Their "Dollars for Docs" database, as they call it, was a veritable story engine. Something as simple as sorting the database and listing which doctors received the most pay became the core of a series of excellent features.

Another way to extract stories from a large data set is filtration: selecting a subset of the data to analyze. Filters can be easy to do, such as when the *Chicago Tribune* extracted Illinois physicians from the Dollars for Docs database for their own analysis. Or they can be much more ambitious, such as when the *Atlanta Journal-Constitution* gathered more than 100,000 medical board disciplinary documents and, from this data set, extracted roughly 6,000 documents that detailed physicians who were disciplined for sexual misconduct. In this case, the newspaper used machine learning—a program trained to recognize phrases related to sexual misconduct—to do the filtration for them.

Finding patterns in data—looking for trends, examining regional or temporal differences, spotting clusters or outliers—not only can lead to a story on its own, but also provides an easier way to find an entry point into a big, complex narrative. When reporters from the *Charleston (W. Va.) Gazette-Mail* wanted to write a story about the opioid crisis in rural communities, they used Drug Enforcement Administration data to track where pain medications were going. When they discovered that more than 9 million hydrocodone pills had gone to the town of Kermit, population 392, it became a natural focus of the narrative. (The resulting series won a Pulitzer Prize in 2017 for investigative reporting.)

Perhaps the most powerful thing to do with data sets is to weld them together and make them more useful than the sum of their parts. It's not always easy to do, but when you bring several different data sets to bear on the same entities, you can extract information that you couldn't get otherwise. For example, the NIH RePORTER database lists all the grants given out by the institutes in the past few years. Combining it with the Dollars for Docs database generates a list of all the researchers who have taken pharmaceutical money while doing federally funded research. All of a sudden, you've got a database that's a treasure trove of financial conflicts of interest in federal research and potential violations of federal law. Similarly, joining a database of prescriptions written for Medicare patients with the Dollars for Docs data set yields a collection of records that reveals how successful pharmaceutical companies are at using money to encourage physicians to prescribe their drugs.

There are lots of ways to get story ideas out of data, story ideas that are often completely novel. When you clean a data set, transform it, display it, filter it, or join it, you are likely becoming the first explorer in uncharted territory. You've automatically found your way out of the pack.

Tools to Handle Data

My first step with a new data set is often to put it (or a portion of it if it's really large) in a spreadsheet program like Microsoft Excel. There are lots of more sophisticated and specialized programs out there for processing data, but spreadsheets are great for a quick initial look. Excel has a feature known as pivot tables that are great for this.

A pivot table is little more than a handy way of aggregating data into tables. For example, you might have a spreadsheet listing all unprovoked shark attacks in U.S. waters. (Yes, you can find several places on the web to get that spreadsheet!) If you put this list in a pivot table, you can group and slice this list in various ways with a minimum of fuss: You can generate a list of attacks by state, by year, or by whether or not the attack was fatal—and switch back and forth between them with just a click or two of the mouse.

With pivot tables, you can quickly come to understand even a very complex multidimensional data set and get a sense of what the fields contain and how they relate to one another. In fact, Microsoft Excel might be all you need for your data analysis. It has some ability to graph data and a surprisingly sophisticated statistical and data analysis package. It tends to choke on large data sets and has a nasty habit of subtly altering certain types of data files if you're unwary, but it's just fine for most purposes.

Graphing or plotting data is another great way to get a handle on the stories that the data are trying to tell you. Unfortunately, Excel's ability to visualize data is fairly rudimentary. With some practice, you can get some reasonably nice-looking graphs, but for serious visualization, journalists tend to turn to specialist applications. Tableau, for example, is a fairly popular visualization program and with minimal tweaking can produce high-quality graphs and maps.

Mapping geographic data is a whole subspecialty of its own, and you might find yourself using a geographic information system (GIS) to build data maps. The industry standard, ArcGIS, is (pricey); a free alternative is Quantum GIS. There are oodles of GIS-compatible maps and other resources that are also free to download.

For big and complicated data sets, or for intricate visualizations, I tend to pull out the big guns: writing my own computer programs and scripts. I favor Python, which is popular, flexible, and well suited for data scraping; its biggest downside is that it can be slower to execute than other programming languages. Some other favorites in the data journalism community are the R language, which is good for statistics, data manipulation, and graphing; and JavaScript, a programming language that's superlative for building applications and graphics on the web.

The great thing about learning a programming language is that it can take care of all of your data needs in one big bundle. I've used Python to scrape data from websites, to request data sets from APIs, to filter huge volumes of information to extract what I need, and to analyze that information. I've used Python packages to plot and map the data I've gathered—and to put data sets into a nice format so I can share them with others. I've even used it to control data-gathering instruments directly: I once used a Python script to log data from an antenna that I had set up to monitor airplane traffic flying overhead for a story I was working on about aerial surveillance. (And less journalistically interesting but no less satisfying: I used a script to control a camera to catch someone who was stealing my morning newspaper!)

Data in Your Story

In some ways, a data set is like any other journalistic source: If it tells a compelling enough tale, it can be the centerpiece of your story. But it seldom will be able to stand on its own; you'll have to use other sources to help interpret what it's saying, to put it in the proper context, and to give you confidence that your conclusions are robust enough to survive the inevitable errors and glitches that creep into every database.

Just like human sources, data sources are most useful when you've got an angle in mind—questions you'd like your data set to answer or a gap in your knowledge that you'd like it to fill. Interrogate it with a sense of purpose, and it'll drive you more directly to a story than if you noodle around and hope for an idea to emerge.

While some sources are clear and straight to the point, some are rambling and prolix, and it's your job to whittle down the extraneous matter to get to the core of the story underneath. This is true of visualizations, too; most good visualizations convey one or two ideas really well. Three or four are probably overdoing it—and risk becoming decoration rather than an illustration that advances the story.

In some cases, the process of gathering, refining, and analyzing data might itself be interesting enough to merit a few paragraphs or even a sidebar. It's usually in your interest to be as transparent as possible about how you gathered, cleaned, and analyzed the data and perhaps even to make the data set available to others. At the same time, you should be aware that many data sets, especially when joined with other data, can undermine people's privacy. Big data are often made out of little people's secrets.

The Takeaway

One of the things I love about data sets is that they've been a never-ending font of ideas for stories that aren't dependent on scientific journals, conferences, preprints, or scientists' willingness to talk. And in some ways, good data sets are easier to find than reliable human sources. Still, I never would make a data set bear the weight of a story alone any more than I would depend on a single interview for a news story. And though the process of getting a data set to spill its secrets uses a different set of tools than extracting information from interview subjects, I find that the two complement each other beautifully. My human sources often lead me to new data sets—and a good data set can give me information that can sharpen an interview.

You don't need to be a programming guru to do good data journalism: You just need to figure out how to ask the right questions of data in the right ways, just as you've probably already done for your human sources.

12

Opinion Writing

Bina Venkataraman

Bina Venkataraman is the editorial page editor of the Boston Globe *and a fellow at New America. She is the author of* The Optimist's Telescope: Thinking Ahead in a Reckless Age *(Riverhead, 2019), named a top book by the* Financial Times *and a best book of the year by Amazon, Science Friday, and National Public Radio. She formerly served as senior advisor for climate change innovation in the Obama White House and advised the President's Council of Advisors on Science and Technology. Before that, Bina was a science writer for the* New York Times, *the* Boston Globe, *the* Christian Science Monitor, *and other publications. She is a frequent public speaker whose appearances have included the TED mainstage, Aspen Ideas Festival, MSNBC, CNN, and university campuses around the world. Bina teaches in the program on science, technology, and society at the Massachusetts Institute of Technology.*

"Everyone is entitled to his own opinion, but not his own facts," Senator Daniel Patrick Moynihan wrote in a 1983 op-ed in the *Washington Post*, popularizing a sentiment that has been expressed by public figures in politics, media, and finance for at least seven decades. In an era when people are entitled to not only have opinions but also publish them widely on social media, the adage ought to be updated: Everyone is entitled to their own opinions, but the ones worth reading are grounded in rigorous evidence and reported facts.

Put another way, opinion journalism is different from merely having an opinion and expressing it. Opinion journalism requires the same diligence as other forms of reportage, the same consideration of sources and citing of evidence, and the same commitment to finding and telling the truth. But opinion journalism is different from feature journalism or news in important ways. A news article might tell you the planet is warming at an alarming rate. A feature article might describe in vivid detail the effects on the world's coral reefs. An opinion column will tell you what ought to be done about it by world leaders or scuba divers—or it might argue that the climate crisis tells us something new about humanity or our history.

Bina Venkataraman, *Opinion Writing* In: *A Tactical Guide to Science Journalism.* Edited by: Deborah Blum, Ashley Smart, and Tom Zeller Jr., Oxford University Press. © Oxford University Press 2022. DOI: 10.1093/oso/9780197551509.003.0013

Unlike straight news or feature writing about science, opinion journalism by definition requires the writer to take a point of view on a matter of ethics, politics, policy, or the public interest and to convince the audience that this point of view is worth considering. Opinion journalism is an invitation to be transparent about one's beliefs and conclusions and, importantly, the reasoning, evidence, and personal experiences that led the writer to come to them.

To write a compelling piece of opinion commentary, whether in the form of an op-ed, an editorial, a podcast, or a book, is to engage in the art of persuasion. While a feature article on science might have the goal of provoking wonder or delight and a news article has the goal of informing, an opinion column or polemic book not only can do those things but also carries the explicit goal of provoking a reevaluation of a belief or viewpoint, or at least a response of agreement or disagreement, from its readers. The best opinion journalism, in my view, provokes readers to see the world in a new way, galvanizes them to action or causes them to question their prior views on a subject of importance to their lives or to society.

Types of Science Opinion Journalism

As an opinion editor at a digital and print news publication, when I consider whether to publish a piece of opinion journalism, I look to see if it offers at least most of the following: a surprising point of view that its writer is well positioned to argue, a robustly defended policy recommendation or call to action that has not already been widely discussed in the public sphere, an unconventional way of looking at a problem that points to new solutions, the refutation of conventional wisdom on a topic of wide relevance, and/or salience to the current news cycle and cultural and political moment.

Opinion journalism that deals with science typically takes one of three forms today:

1. Science as the font of facts. Science is used as a source of evidence in an argument made in an opinion column or book about a matter of public policy, politics, ethics, culture, or personal decision-making. For example, a pair of doctors argue, as they did in the pages of the *Boston Globe* opinion section in April 2020, that, contrary to what the Centers for Disease Control and Prevention was then recommending, everyday people, not just medical workers, ought to be wearing masks in public. To make this argument, they drew on the science that suggested that

Covid-19 was being spread primarily in the air, citing research studies and their own expertise.

2. Science as the target or terrain of critique. An opinion commentary reflects on something happening in the realm of science and technology, critiquing a practice or approach, often with the aim of ensuring that research and innovation better serve the interest of society. For example, the *Boston Globe* editorial board in January 2021 argued that rampant disinformation and hate speech on social media, paired with business models that tied advertising revenue to virality of false or inflammatory content, showed the need for stronger regulation and alternative social media funded by philanthropy or public sources. In doing so, the board evaluated the ways that algorithms lead people down rabbit holes of conspiracy theories and hate and the inadequacy of technology companies' efforts to tame these threats to democracy and civic discourse.

3. Scientific inquiry as a value. An opinion writer takes a position on a matter of policy, ethics, or politics driven by the belief that the scientific enterprise and scientific integrity are important to society. For example, in 2020, *Scientific American* and *Nature*—the former for the first time in its history—endorsed a candidate in the U.S. presidential election, citing the need for policies grounded in science to address grave threats posed by the pandemic and climate change, and the need to protect the ability of scientists to do independent research that guides the public and policymakers. In doing so, the publications' editorial teams argued that protecting science from political interference had the effect of serving the greater public good.

Reporting and Refining a Point of View

An opinion writer may start out with a point of view before writing a column. (For example, climate change ought to be addressed by governments, not just businesses.) Or the writer might just start from a place of curiosity about a topic of relevance: Should cities allow indoor dining during an outbreak of a rapidly spreading airborne virus? The process of reporting and deciding on a perspective or position is ideally an iterative one; in reporting, a writer may find that her original position and point of view change entirely. At a minimum, it ought to be further refined; if not, the writer should question whether she has done enough reporting and evaluation of the facts and counterarguments.

Either way, the writer should be open to what reporting shows about the topic and her opinion. Tacking back and forth between digging for evidence and orienting toward a surprising or unique point of view makes for more cogent and evidence-driven opinion journalism—the kind that distinguishes itself from a tweet or Reddit post that anyone could post without doing their homework.

The seed for a great opinion column or polemic book is often a question, not a foregone conclusion. To put a finer point on it, science should not be wielded as a sword to prove one's existing beliefs. Evidence that doesn't support a writer's "take" or position should be engaged with directly, not cast aside. Correlations between phenomena (people who drink red wine also happen to live longer) should not be automatically assumed to be causations (drinking red wine extends your life span or old people crave more red wine). Unfortunately, it's all too common to encounter op-eds and books that dangerously mislead the public by cherry-picking data or studies to prove a point the writer was dead set to make, regardless of the evidence. I consider such practice to be propaganda, not opinion journalism.

Good science journalism requires gauging what the evidence in totality shows and looking for areas of scientific consensus, not isolated studies, to bolster a point of view. Peer-reviewed science should be privileged over hunches from people who happen to have PhDs, and even published journal articles should be vetted with expert sources who might see the gaps in the methodology or the blind spots of the experimenters.

The typical rules of journalism apply to science journalism and to opinion journalism on science, in that the writer should be skeptical and critically evaluate any information received. At the same time, writers should be cautious about being or amplifying lone voices criticizing the preponderance of experts: That is, a political columnist who is not a scientist ought to think hard before disagreeing with tens of thousands of climate scientists whose research over decades shows that atmospheric carbon dioxide is rising and warming the planet. That said, the columnist might disagree about the political solutions to that phenomenon and might argue, for example, that it's city and state governments, not the federal government, that ought to be responsible for solving the problem. (In doing so, the columnist ought to be prepared to answer the counterarguments that climate change is a global problem and that state and local governments are often strapped for cash.)

Coming up with a clear point of view for an opinion piece often requires balancing what evidence or research shows with what it means for society. The excavation of scientific evidence for an opinion column must be paired with a consideration of competing values at play. For example, using the example above on indoor dining, many questions can be asked: How much is known about the risks? What is the cost of closures to small businesses? What will happen to restaurant workers? Are there ways of resolving the trade-offs, for example, through better aid to restaurants and restaurant workers to offset indoor dining bans?

Some of these questions are a matter of scientific fact, but some are a question of weighing the trade-offs. The way to answer these questions involves a combination of interviewing experts, consulting scientific journal articles, researching a variety of points of view, and thinking through the best arguments that support various positions. As the writer reports and weighs the different considerations and values at play, the refines his or her point of view. Ideally, in the process, the most important counterarguments to the writer's point of view, which ought to be addressed in the piece she's writing, are refined.

The Art of Persuasion

When I've taught college and graduate students how to write effective op-eds, I've often called on Aristotle's three methods of persuasion, as outlined in his work *Rhetoric*: ethos, logos, and pathos.

Ethos is about establishing the credibility of the speaker or writer who is trying to persuade. For a journalist or expert writing a piece of opinion journalism about science, this can mean establishing that you have expertise, good intentions, and even the kind of character that makes you a credible authority on the subject. I try to remind people that expertise can come from not only education, but also personal experience. For example, a contributor to the *Boston Globe* opinion section is a grocery store clerk, Mary Ann D'Urso, who offered perspectives on essential workers during the pandemic.

Logos involves using the tools of logic and evidence to persuade readers. That means using data or scientific studies; citing or quoting experts or authorities on a matter of dispute; drawing on analogies, precedents, or supporting examples; and making plain the logical reasoning that led to your conclusions. For good opinion journalism on science-related themes or that uses science as a source, it's important to draw on multiple tools of logos.

Pathos is persuasion by emotional appeal. Eliciting an emotional response from your readers can lead them to share your point of view and can be done by imparting stories of people in struggle or triumph and by using powerful and rousing language. Descriptions of scenes or of people whose circumstances create a natural emotional reaction can also serve the writer well. For many audiences, facts and logic alone are not enough to persuade them to take action or to change an existing viewpoint. An emotional response, for example, to a family that has suffered the consequences of a flood, however, might change their perspective on the risks of flooding. Don't underestimate the persuasive power of emotional appeals or how emotions might create immunity on the part of readers to logical arguments.

Key Tips for Opinion Writers

Use facts as scaffolding, not ammunition. Don't barrage people with data points or scientific facts as a way of persuading them. Recognize people respond to pathos, logos, and ethos—show them your credibility, which involves becoming trusted as a writer over time and within the context of each piece you write. Don't discount emotional appeals; do focus on solid reasoning and logic paired with selective evidence.

There should be no ostriches in opinion writing. One common pitfall of novice opinion writing is pretending that good counterarguments to your point of view don't exist. It will make your opinion piece stronger if you acknowledge what will be readers' natural sources of skepticism and outline why these concerns failed to sway you; this will build further trust with the reader rather than create the impression that you failed to consider other points of view. Take on the best or most widely held counterarguments and argue why your point of view is still worth having in spite of them.

Don't cherry-pick data points or scientific studies. Do evaluate bodies of research, consult a range of credible experts, look for recent and peer-reviewed work, and talk to people who might disagree with the conclusions you or the scientists you are consulting are drawing.

Balance what you learn about the science with a human-centered perspective. Don't presume that everyone trusts scientists or experts—or that the science speaks for itself. Do speak in terms of people's values and recognize that science might suggest one course of action but that morality or politics might suggest a different course. "Listening to the science" doesn't usually yield a simple answer to questions that involve real people and real trade-offs.

The Takeaway

Opinion journalism plays a powerful role in shaping public dialogues and decisions. It's made more rigorous when journalists, authors, and op-ed writers draw on credible scientific research—while weighing that evidence alongside the competing values at play in any political or personal decision. Society also benefits from opinion writing that challenges scientists and technology innovators to better serve the public interest. The perspective that a writer takes ought to be shaped and refined by reporting in an iterative process.

Great opinion writing provokes readers to think in new ways and to question or affirm their own views on a subject. It should strive to persuade a broad audience to consider the writer's perspective by drawing on all three persuasive modes: ethos, logos, and pathos.

Persuasion, the ultimate goal of opinion writing, is best achieved by engaging with strong counterarguments and by transparently wrestling with the values and trade-offs inherent in taking the writer's position. Opinions might as well grow on trees in the 21st-century media ecosystem rife with Twitter tirades and broadcast punditry. But good opinion journalism backed by rigorous reporting and thoughtful analysis still rises above the rest; it is a meaningful endeavor for any writer and a great public service.

13

Magazine Writing

Paige Williams

Paige Williams is a staff writer at the New Yorker *and the author of* The Dinosaur Artist, *which the* New York Times *named as one of its 100 Notable Books of 2018. Her journalism has won the National Magazine Award for feature writing and has been anthologized in* The Best American Magazine Writing *and* The Best American Crime Writing. *She was the Laventhol/Newsday Visiting Professor at Columbia University's Graduate School of Journalism and has taught at a number of other universities. She has been a fellow at MacDowell and was a Nieman Fellow at Harvard University.*

In May 2012, I came across a science story that presented as a crime narrative. It involved the auction of a very large, very illicit *Tarbosaurus bataar* skeleton from the Gobi Desert of Mongolia. The Gobi is unusually rich in the fossil remains of Late Cretaceous creatures and plants, and although Mongolian law prevents individuals from collecting or selling them, dinosaur body parts had been pouring out of the country and to the commercial market for years. The *T. bataar* skeleton, which represented what has been called the Asian twin of *Tyrannosaurus rex*, surfaced at a fine-art auction. The sight of a fully assembled, well-prepared apex predator could not be ignored. Lawyers and politicians tried to intervene, but the skeleton sold, to a mystery buyer, for over $1 million. The transaction was immediately placed on hold until questions of provenance and ownership could be unraveled.

The auction alone was a contained narrative: beginning, middle, and end. The event was covered in the daily news and then dropped, as happens so often. To me, the sale only raised more questions. I sensed the potential for a deeper story because I had been obsessively studying the tension between paleontology and the black market for several years, looking for a way "in." I already knew that fossil hunters often called themselves "commercial paleontologists"—which are not a thing. A paleontologist is a paleontologist; paleontologists do not sell fossils or condone the commercialization of the

Paige Williams, *Magazine Writing* In: *A Tactical Guide to Science Journalism.* Edited by: Deborah Blum, Ashley Smart, and Tom Zeller Jr., Oxford University Press. © Oxford University Press 2022. DOI: 10.1093/oso/9780197551509.003.0014 objects fundamental to their research. Yet this was not to say that scientists were inherently the best at finding, excavating, and preparing fossils.

Long-form magazine journalism allows reporters to tease out the nuanced, complex interplay of cultural forces and characters. A story may take one of various forms: narrative, essay, explanatory reporting, or a combination thereof. Every piece, regardless of genre and no matter how daunting, begins with fundamental questions. It was impossible to report and write "Bones of Contention," published in the *New Yorker* in January 2013, without beginning the inquiry with the mystery seller and buyer. Who would go to the trouble of acquiring fossil dinosaurs from Mongolia? How did those bones get from the Gobi to New York City? Who knew how to prepare and mount such a nice skeleton? Who buys dinosaurs? Most crucially: What does science lose when clues about the history of life on Earth vanish into private collections?

The questions led to a surprising mix of answers involving paleontology, collectors, the fine-art market, property laws, opportunism, subterfuge, ambition, marriage, cultural conflict, post-Communist Mongolia, and diplomatic relations. (I certainly never expected to find the ghost of Newt Gingrich in Mongolia.)

A story that zinged from Gainesville to Manhattan to Ulaanbaatar to Munich to Tucson to Hollywood, and to points in between, allowed for variation on characters and setting (i.e., texture). These promising clues had been embedded in the auction. The seller turned out to be a fossil dealer with an engineering degree, a wife, two small children, a history of quiet rebellion, and a talent for historical salvage. As I was reporting the story, federal agents arrested him on charges related to smuggling, which raised a whole new set of questions, especially those involving why the Department of Homeland Security would get involved.

I give you this backstory only to explain that I don't know of a good long-form magazine project that did not start in obsession and that is not executed with careful attention to reporting, above all. Authentic curiosity is Step 1.

Find the Tension

Magazine features can fail at the idea level when the reporter struggles to discern topic from story. Paleontology is a topic. The tension between paleontologists and commercial fossil hunters also is a topic. A Florida man's strange journey from obscure collector to accused international dinosaur smuggler, that's a story. More specifically, it is narrative. Each character had a goal, met resistance, and saw resolution.

Still, it's important to remember the bigger picture—the context—and to capture both the fresh *and* the universal. Plenty of people have written about Florida as a natural wonderland or as an overdeveloped hellscape. In "Orchid Fever," the 1995 magazine piece that became the masterful book *The Orchid Thief,* Susan Orlean managed to do both via a lyrical factual study of an eccentric man who sloshed into swamps to poach rare plants. Orlean used a federal court case and a fascinating character to show us a new way of looking at botany, psychology, ecology—of humankind's relationship to an imperiled environment.

One caution about the importance of staying nimble as the reporting unfolds: Never try to "fit" your story to an underinformed conceit. Editors often conceptualize stories and assign them in good faith, but the reporting must bear out the idea.

Stay Organized

Long-form magazine writing can feel like a million-part process. Find the story, report the story, understand the story, re-report, revise, edit, revise again (and perhaps again and again), and then fact-check. (Fact checking, to me, is satisfying and fun.)

I've always thought it a mistake to think of these as discrete acts. The reporting shouldn't necessarily stop after the writing begins, especially if you're working on time-sensitive material. And the fact check should be fundamental to the reporting and writing. It amazes me to hear a writer say, "I'll just leave this for fact checking." If you don't know what's factual, how do you even know what to write? The reporter is the front line of defense.

In long-term reporting projects, information branches off, circles back, dovetails, and dead ends. Close your mind and you may miss something, but the more information you gather, the better organized and deliberate you'd better be about managing it. You'll need to know *how* you know what you know. It takes far less time to develop an index or a timeline than it does to backtrack, in a mad scramble.

Be meticulous about sourcing. On Twitter and elsewhere you may have seen people discussing methodology: index cards, DEVONthink, Scrivener, whatever. I used to fetishize the gadgetry part of the process but decided that this was a form of procrastination. Find what's best *for you* and stick to it for as long as it works.

Build It Well

Structure, as a component of craft, doesn't get enough attention. It is often overshadowed by a more obvious element, voice. If you want people to read your work you have to be strategic about holding their attention: Long-form pieces run from 3,000 to 10,000 words or more. Magazine writers are blessedly free from the tyranny of the nutgraf, the nugget of summary that anchors daily newspaper articles. Instead of training a harsh spotlight on timeliness and takeaway, the magazine writer may pan the landscape, asking the reader to settle in for an experience or for a slower release understanding of an issue, event, or person. As the experience's architect, you must build a house that people want to live in.

Students of magazine writing know that long features are built section by section. Each section has a purpose. When deciding on a structure, the writer is thinking about a range of factors: complexity, assigned length, characters, and deadline. If you autopsy canonical magazine writing, ask yourself how each piece is built. What happens in each section? How many sections are there? How many words in each section? How many characters in each section? What is the narrative purpose of each section? Is there dialogue? (Dialogue is different from quotations.) Did the author witness the dialogue or find it, say, in a court transcript?

As you're charting the reporting behind a favorite piece of writing, it can be useful to think about the likely source of each bit of information—human sources, documents, or personal observation.

How does the writer begin? With an expositional passage? In media res? With a declaration? A physical description of the main character? A description of a setting?

Chronology is the bedrock of structure. How can you understand any story that you are writing, especially a complex one, unless you know the order in which events happened? Unless you've already got a John McPhee–like ability to intuit or devise a *sustainably* baroque story structure, it is useful to start in ordered clarity and then work your way up to improvisation.

The writer who is thinking in terms of gimmicks won't ultimately get very far. If you're thinking about how you'll cleverly use flashbacks before you've mastered chronology—well, godspeed. "Bones" had to begin with the mystery of the skeleton itself, standing behind velvet showroom ropes, on auction day. Structurally, the first section had to end with an action that left *answerable* questions in the mind of the reader. To shift metaphors, you're a mapmaker, charting a course.

Make Me See

Visuals matter for the same reason structure matters: engagement. I'd argue that visuals matter *especially* in science writing, which requires a clear conveyance of often-complex ideas. It helps to have a mind for metaphor. (McPhee's description of Deep Time is the standard bearer: "Consider the earth's history as the old measure of the English yard, the distance from the king's nose to the tip of his outstretched hand. One stroke of a nail file on his middle finger erases human history.")

If you don't notice something, you can't describe it. Visuals begin in reporting, and reporting demands perception. Reading fiction and poetry opens the mind to ways of seeing. A writer's notebook should be filled with detail and description, even knowing that you will almost always need to winnow the material to a single, vivid image. Ideally, a description conveys more than the nuts and bolts of, say, dimension or color, and at least acknowledges Chekhov's standard: "Don't tell me the moon is shining; show me the glint of light on broken glass."

In "Bones," it was important to not only describe the skeleton but also contextualize it as an ancient apex predator caught in an almost comically bizarre modern context: "Eight feet tall and twenty-four feet long, the specimen had been mounted in a predatory running position, with its arms out and its jaws open, as if determined to eat Lot No. 49220—a cast Komodo dragon, crouching ten yards away, on blue velvet."

Tarbosaurs did not eat Komodo dragons, but the juxtaposition allowed worlds to collide, and it conjured the food chain—literal and figurative. I did not have to work hard for this: The tension was standing there in plain sight. The fact that I had not attended the auction was no hindrance: I was able to describe the scene by triangulating information from the auction catalogue, videos, photos, and interviews.

The Sound on the Page

As a responsible and credible magazine journalist, you don't monkey around with fact, but you can be creative at the structural level and with voice. Voice is *how* the story is told—pitch, tone.

I've often asked students to analyze a magazine piece for voice and describe it in a single word. They may say *somber, chatty,* or *wry.* Joan Didion does not sound like David Foster Wallace. The writer can calibrate the reader's experience through subtle shifts in sentence length or word choice or, if the publication allows it, punctuation. Didion, who writes with a catlike aloofness, keeps us at arm's length. Ed Yong wrote with conversational urgency about the Covid-19 pandemic: "In October, the Johns Hopkins Center for Health Security war-gamed what might happen if a new coronavirus swept the globe. And then one did. Hypotheticals became reality. 'What if?' became 'Now what?' So, now what?"

Voice can shrink the distance between writer and reader. Didion's first-person elegance almost suggests that she'd rather go unnoticed. Yong's voice is innately witty and charming. Wallace's maximalism shakes the reader by the scruff. If you've ever played outdoor tennis in the Midwest it may never have occurred to you, before reading Wallace, that you faced a shadow opponent—a savage wind. In his sonic masterpiece "Derivative Sport in Tornado Alley," which ran in *Harper's* in 1992, Wallace managed to capture the converging forces of math, physics, entomology, geology, biology/agriculture science, and meteorology by explaining that the residents of his native Philo, Illinois, knew that "to the west, between us and the Rockies, there is basically nothing tall, and that weird zephyrs and stirs joined breezes and gusts and thermals and downdrafts and whatever out over Nebraska and Kansas and moved east like streams into rivers and jets and military fronts that gathered like avalanches and roared in reverse down pioneer oxtrails, toward our own personal unsheltered asses."

Don't try to do this—the tornado tennis or the writing. There will never be another David Foster Wallace, just as there will never be another Joan Didion. Find your own way. Authenticity is never gimmicky.

Find Your True North

What are you trying to do, as a writer? It helps to have a clear goal, which doesn't mean having an agenda. Alerting the public can be the highest form of service; entertaining them along the way is art. In "The Really Big One," Kathryn Schulz described—with an improbably successful combination of horror and levity—the earthquake that will devastate the Pacific Northwest. The indelible images include a grown man trying to decide whether he could fit beneath a school desk and a refrigerator unplugging itself and marching across a room. Schulz's piece pricked the public consciousness in an urgent new way. Yong's work did the same as the Covid-19 pandemic unfolded, providing the explanatory framework for a disaster that we were all trying to survive, in real time.

Not every piece or body of work can have that sort of impact, but you can always choose to take the work seriously (without taking yourself too seriously). You owe no one an explanation of your motivations. When I was working on the dinosaur project, one friend lamented that I seemed unable to get interested in "things that are alive"; another questioned the value of a "niche" story. I saw it another way. Magazine writing offers the potential for not only for depth but also tapestry. It's fun, weaving together worlds and making connections that did not previously exist.

The Takeaway

Magazines inherently offer certain freedoms that other journalistic forms do not, namely, the space and time to go deep or develop characters. These freedoms mean nothing unless you learn how to control them. If a magazine demands a certain house style, figure out a way to work within it, in your own way. Know your purpose. Are you trying to serve the public, to entertain, neither, or both?

The best teachers are practice and study. Build a personal canon, with range—Wallace here, "Hiroshima" there. Read obsessively in a range of genres. Find the right guides. *The Open Notebook* is an invaluable source of craft talk, and you can also learn a lot by studying the works collected in *The Best American Magazine Writing* and *The Best American Science and Nature Writing.*

Reporting is everything: make the calls, get on the ground, vet the studies, and find the documents. Writing nice sentences is never enough. Strong reporting plus lovely writing: These are a superpower.

14

Book Writing

Dan Fagin

Dan Fagin is a professor of journalism at New York University, where he directs the Science, Health, and Environmental Reporting Program and the Science Communication Workshops. He has taught more than 300 science journalists, about 20 of whom have written books so far. Dan is the author of the New York Times *bestseller* Toms River: A Story of Science and Salvation *(2013), which was awarded the Pulitzer Prize for General Nonfiction, the National Academies Communication Award, the Helen Bernstein Book Award for Excellence in Journalism, and the Society of Environmental Journalists Rachel Carson Book Award, among other prizes. Earlier, Dan spent 14 years as the environmental writer for* Newsday, *where he was a principal member of two reporting teams that were Pulitzer finalists. A former president of the Society of Environmental Journalists, he is also the coauthor of* Toxic Deception *(1997), which was a finalist for the Investigative Reporters and Editors book of the year award. Dan's next book is about monarch butterflies and the future of life. It's taking forever, but he'll get to the finish line eventually!*

I've never run a marathon, but I feel like I have because I've written a book. I'm working on my third and can testify with confidence that a book project will challenge you like nothing else you will encounter during your career as a science journalist.

Books require more of everything: more planning, more pitching, more research, more travel, more interviews, more outlining, more writing, more rewriting, and then *more* rewriting. Book projects generate more frustration and more stress, too, but in the end, they will also reward you in ways nothing else in journalism can. If your book turns out well, you will reap benefits that will boost your career, even if (like most authors) you don't make much money from the book itself.

There are psychic rewards, too, from knowing you've met the ultimate challenge in our profession and made a lasting contribution to the topic you have explored so thoroughly.

Dan Fagin, *Book Writing* In: *A Tactical Guide to Science Journalism.* Edited by: Deborah Blum, Ashley Smart, and Tom Zeller Jr., Oxford University Press. © Oxford University Press 2022. DOI: 10.1093/oso/9780197551509.003.0015

Even in an era when daily journalism can live online for years, there is a sense of *permanence* to books that is unique. Whether you sell a few thousand copies or a hundred thousand (I've done both), people you've never met will be reading, citing, and talking about your book for years, which is a great feeling!

My first rule of book writing is: Don't start a project you're feeling ambivalent about. To put it bluntly, you'd better love your book idea at the beginning because you'll probably hate it at the end—until it's finally out of your head and the finished product is in your hands and the hands of your readers. That's the moment when you can finally appreciate your book for the remarkable, if exhausting, accomplishment it is.

A quick caveat before we continue: The advice in this chapter is tailored to one kind of book: commercially published nonfiction narrative. There are, of course, markets for other kinds of science-related books, including textbooks, how-to guides (like this one), popular explainers, science fiction, academic discourse, and children's books. Each of these genres has its own tips and traps. This chapter focuses on the specific type of science book that stands the best chance of finding a large audience: the kind that tells a deeply reported story.

Like a marathon, a narrative book project is something you should pursue in stages. There's an off-ramp at each stage, but the farther along you get, the harder it is—emotionally and financially—to stop running. The key, as with *everything* in journalism, is to take those early steps seriously. The more attention you pay to finding a great idea and making a great plan, the smoother the rest of the run will go. So let's start at the beginning, with idea formation.

The All-Important Central Idea

You know what a book is in a physical sense: a bound aggregation of pages or a digital facsimile thereof. But even more than that, a nonfiction narrative book is an *extended idea*. Elizabeth Kolbert, one of our finest environmental writers, spent most of her career writing newspaper stories and then magazine stories. Her first book was a collection of pieces she had written for the *New Yorker* about New York City politics: *The Prophet of Love and Other Tales of Power and Deceit* (2004). Those stories were excellent on their own, but there was no unifying theme, no arc, no journey—just a series of insightful but disconnected profiles of city politicians.

Her next book, *Field Notes From a Catastrophe* (2006), was also based on her magazine reporting, about climate change, but this time she did more adding and rewriting to make the overarching theme more explicit. Her two last books, the Pulitzer Prize–winning *The Sixth Extinction: An Unnatural History* (2014) and *Under a White Sky* (2021) were conceived as cohesive, full-length narratives and read that way, even if some portions were published first in the *New Yorker*. Those last three books, each more focused than the one before, solidified Kolbert's place in the highest echelon of our profession. The lesson here is clear: Books work best when they consist of *one idea, extended and explored*, instead of multiple ideas mashed together.

So, where can you find a book idea that works? You can start by simply being a busy journalist and wait for a sufficiently compelling idea to arise from your daily work. That's much more likely to yield a book idea original enough to stand out from the crowd of would-be authors clamoring for the attention of agents and publishers. Remember, too, that you're more likely to catch their attention if you're seen as someone with a track record of success in producing shorter narratives—especially if the topics are similar to your book idea. In a nutshell: Don't try to conjure your book idea out of thin air; wait for it to arise from your daily work instead.

An Agent, a Proposal, and a Publisher

Getting an agent is not an absolute must, but most authors will tell you that you should if you hope to reach a large audience. While it's certainly possible to deal with publishers directly (especially academic and small houses) or to self-publish (more feasible than ever in the digital era), there are big advantages to representation. Few books can find a large audience without a commercial publisher, and few books are commercially published without a literary agent. The only downside is that an agent will take 15% of your book earnings.

Is it worth it? Almost always. A good agent is your adviser as you shape your proposal and your advocate as you negotiate with publishers. They are your envoy to an opaque and idiosyncratic industry that is often incomprehensible to outsiders. Perhaps most importantly, their involvement confers credibility by signaling to publishers that you and your idea are worth their serious consideration.

If you're a working journalist, you may have already had the experience of an agent contacting you and encouraging you to turn something you've published into a book proposal. If that happens to you, my advice is to be careful. An agent who already loves your work might not be the tough critic you need.

As an alternative, ask for recommendations from successful author acquaintances if you're lucky enough to have them. If you don't, you can consult reference books and online resources such as publishersmarketplace.com, though I suggest grabbing a notebook and visiting one of those old-fashioned brick-and-mortar structures known as bookstores. Make a beeline for the section with books similar to yours and pull out the ones you admire most. Look in the acknowledgments section of each book; almost always, the author will thank their agent. If you look inside enough books, you'll start to see familiar names popping up.

Reach out to prospective agents with a short email to introduce yourself, summarize your qualifications, and explain your idea in a few well-chosen paragraphs. If the agent is interested, they'll reach back to you and set up a meeting, at which point you can size them up. Does the agent have a successful track record in your genre? Do they have insightful comments and suggestions about your idea? Will they give you sufficient time and attention? Do you *like* them? (Personal chemistry counts for a lot in publishing.) If those answers come up "Yes," you'll sign a representation agreement. Agent contracts tend to be standardized, but it can't hurt to have a lawyer familiar with the publishing industry take a look before you sign.

Now it's time for the actual book proposal. This can be as short as a half-dozen pages or as long as 100 pages, but all start with a short overview of the project, which includes a marketing plan identifying your audience(s) and how you hope to reach them, usually followed by a chapter-by-chapter outline (with a few choice sentences about each chapter) and an author biography. For the marketing section, you'll also include information about your own contacts (including prospective blurb writers) and your social media footprint. Finally, your agent will probably want you to include a sample chapter that shows off your writing and reporting ability. If you're a journalist who has already produced stories on the topic, you may want to include links to them in your proposal. Writing an effective proposal can take months, so don't rush the process—it's not only crucial to selling the book, but also will help you later when you're organizing the book itself.

Your agent will take the lead by sending your proposal to one or (more likely) several publishers, who will then decide whether they want your book and how large an advance they're willing to pay for it. An "advance" is just that: an advance payment against your per copy sales royalties. For example, an author might get a $60,000 advance split into thirds: $20,000 when the contract is signed, $20,000 when the finished manuscript is accepted by the publisher, and the final $20,000 on publication. If the book ultimately sells enough for the author to "earn out" the advance (only about one in four books do), they will then start getting royalties on additional sales. Royalties range from 7.5% to 15% of the retail price of each book sold, depending on the format. Royalties for an e-book are usually 25% of the net digital price.

If you're fortunate enough to have more than one publisher interested in your book proposal, consider more than just the size of the advance. Think carefully about who will be editing the book and who will be marketing it. Do they have good track records and good ideas? Do you *like* them? (Chemistry again!) There are so many factors to consider that in the end, you might not choose the publisher that offers you the biggest advance.

Writing on the Go, as You Go

There are so many ways to write a book that few universal rules apply, but just about every author will also tell you it's important to *write as you go*. There are many reasons not to postpone writing until you're done with your research. You may *think* you know what's going in all of your chapters, but trust me, you don't—not until you write them! Writing chapter drafts is a great way to identify reporting holes. It is also, for many of us, the most difficult part of the process, so don't put off the pain—spread it out instead.

It is hoped, you'll be encouraged by each page you write ("That's one page down; only 249 to go!"). You're a marathoner now, and you need to keep yourself psyched up and feeling good about your progress.

Keep in mind, too, that most books are multiyear projects, which means they will test your organizational skills and your memory. You will need a coherent filing system, whether digital, physical, or (most likely) both. Clever software packages (Scrivener is the best known) can help you with this, or you can rely on Microsoft Word and old-school metal filing cabinets like me. Whatever system you choose, you're going to need to stick with it, and it's going to have to work well enough for you to find the information you need when you need it because you're not going to be able to rely on your memory: Books are just too big for that.

For the same reason, when I'm in the field doing reporting, or even when I'm just interviewing someone over the phone, I make it a priority to set aside some time at the end of the day—no matter how tired I am—to go through my raw notes and write a few coherent paragraphs of key impressions, observations, and quotes. Trust me when I tell you that it's a thousand times easier to decipher your raw notes in the hotel room at the end of the day than 2 years later when you're putting a chapter together.

And I don't only limit my daily organizing to text, either. Digital photos are crucial memory aids. I take dozens of them every day I'm in the field, and I title and file each one at the end of the day so I'll be able to find them easily years later and remember just what that scientist, or that sunset, looked like.

I am equally obsessive about revising my work. Books are so long that even great reporting is not enough to hold your reader. The writing has to be *propulsive*, compelling the reader to keep going, chapter after chapter. That means you'll need to drop lots of breadcrumbs along the way, including transitions and other signaling devices to keep the reader oriented and engaged. It also means that even in a book, you need to be ruthless about cutting passages that are not sufficiently interesting or important. You might think you have more latitude to be verbose in a book, but I think it's the opposite. The more you ask of your readers—and in a book, you're asking a *lot*—the more you need to respect their limited time and attention. If you don't, they'll put your masterpiece right back on the shelf, unread.

Because books are so permanent, it's even more important than usual to be scrupulous about accuracy. Generous footnoting (or more likely, endnoting) of your manuscript helps. So does hiring a qualified fact checker. Ask your author friends for recommendations or head back to the bookstore and read more acknowledgments to see who gets thanked. Unfortunately, you can't count on your publisher to do a thorough vetting. If you're fortunate, though, your book editor will give you lots of wise suggestions—including what to trim. If you're smart, you'll listen. As with any reporter–editor relationship, mutual respect is the key.

It's Not Over When It's Over

In a sense, you've crossed the finish line when your book is done, but not really, because a book, like any product, needs to be publicized and marketed to succeed. It is hoped, your publisher will take the lead, but never forget publishers are a fickle bunch. When things are going well, your publisher is your best friend; when things aren't, the emails and calls stop coming. Don't take it personally, it's just business.

Publishers know they will not make a profit on most of their books. To stay in business, they need to concentrate their marketing and publicity efforts on those titles that stand the best chance of finding large audiences. They will push your book for as long as they think there's a chance it will break through—and not a minute longer. For this reason and others, early positive reviews, including from trade publications like *Kirkus Reviews*, *Publishers Weekly*, *Booklist*, and *Library Journal*, are crucial, as are advance orders from Amazon and other online vendors.

Remember that in the end, no one cares more about your book's success than you do. Even bestselling authors tend to be very involved with publicizing their work, and not merely by showing up for events. They promote their work at every opportunity and actively seek support from their personal networks. Here's where being a good citizen on social media can pay real dividends: If you've said nice things about others' work, they'll gladly return the favor when your time comes.

The Takeaway

In the end, as with so much in life, it's important to manage your expectations when writing a book. Don't expect a bestseller, cherish every positive review, and enjoy the satisfaction and tangible career benefits (like speaking gigs and better assignments) that come with being seen as an authoritative, engaging guide to a field of science. Who knows, someday you might even decide to go back into training and run the marathon all over again.

PART III

INVESTIGATIVE JOURNALISM

15

Investigative Science Journalism

Katherine Eban

Katherine Eban, an investigative journalist, is a contributing editor at Vanity Fair *magazine, an Andrew Carnegie fellow, and author of the* New York Times *bestseller* Bottle of Lies: The Inside Story of the Generic Drug Boom. *Among her many assignments, she investigated the 9/11 hijackers for* The New York Times' *investigative unit and exposed massive data fraud at India's largest generic drug manufacturer for* Fortune *magazine. At* Vanity Fair, *she identified the CIA (Central Intelligence Agency) psychologists who designed the Bush administration's coercive interrogation methods during the war on terror. That article, "Rorschach & Awe," was later made into the film* The Report. *Her work has won numerous awards, including from Investigative Reporters and Editors, the Association of Health Care Journalists, and the Overseas Press Club.*

What exactly is investigative journalism? In its most basic form, it exposes stories or information that powerful interests or individuals want to keep hidden, disclosures that can benefit the public good. Pretty exciting, huh? But the best investigative journalism must go further, not only exposing injustice and malfeasance but also analyzing the systems that give rise to it and the historical patterns that allow it - determinedly tracking who benefits and who suffers.

So how does that work in science journalism? To some extent, it works the same as it does with any other beat. Investigative journalists reveal how malign influences—such as money, power, or the prospect of self-gain—warp the rules, tilt the playing field, and lead to harm.

That scrutiny can become turbocharged in the worlds of science, medicine, and public health, where regulations and oversight are critical, and the stakes can be life and death. You will rarely find yourself thinking that your work doesn't matter. As Dan Diamond, an investigative health journalist formerly at Politico and now at the *Washington Post*, says of covering the SARS-CoV-2 pandemic, "All of society is being remade by the story on your beat."

Katherine Eban, *Investigative Science Journalism* In: *A Tactical Guide to Science Journalism.* Edited by: Deborah Blum, Ashley Smart, and Tom Zeller Jr., Oxford University Press. © Oxford University Press 2022.
DOI: 10.1093/oso/9780197551509.003.0016

Almost every investigative science journalist on the planet, including me, reported on Covid-19. But you don't need to wait for a pandemic to start digging. Wherever there is the prospect of big money or power, there will be rich terrain.

The best stories, however, don't come with a sign that says, "Dig here." There are no eager publicists promoting them. So how do you even know there *is* a story, rather than a dry well that will devour your time and erode your editor's confidence? Believe it or not, the answers lie mostly with some basic tools: your own curiosity, thinking, and dot connecting; well-deployed interpersonal skills; and a refusal to give up.

Identifying the Landscape

So you're at the start of a new story, feeling lost and sweaty-palmed, hoping that a reporting map will drop from your ceiling. When it doesn't, you will need to turn being lost into a virtue. I try to approach each new story as an unfamiliar landscape. My first question is usually, "Where am I?" I try to observe the topography, look around for inhabitants (who can serve as sources), find the watering holes where they might congregate, and note if trails diverge. Your ability to understand *where you are* will impact what kinds of questions you'll ask and to whom you'll ask them.

Here's an example. In 2008, I got a tip from a pharmacologist with a popular radio show: His listeners kept calling in to complain that their generic drugs didn't work. When he relayed these complaints to the Food and Drug Administration, regulators there claimed the reactions were likely psychosomatic, with the patients imagining harm due to a change in pill size or color. But the radio show host didn't believe that and called me instead, posing a question: "What is wrong with the drugs?"

I began reporting from a consumer safety landscape. I found patients suffering from troublesome side effects and doctors puzzled to find that patients they had stabilized became unstable after being switched from a brand name to a generic drug or between different generic versions. My first story was competent enough. I documented the existence of a possible problem. But I was no closer to answering what was *actually wrong*—largely because I was reporting from the wrong landscape.

Over the next 10 years (remember the part about never giving up?), I came to understand that the answers lay in a different landscape entirely. Yes, doctors and patients were the victims. But essentially, they were onlookers. It was whistleblowers inside the generic drug industry who actually had the answers: Companies were faking their quality data in order to get their drugs approved. At its heart, this was a corporate corruption story. Once I figured out the right landscape and started reporting on decisions made by company officials, I nailed the story.

Mapping the Landscape

So you've found the landscape where you need to report. But do you know your way around? Do you understand who the locals are and how they operate? Do you know the history and the rules of the place?

If the idea of a landscape doesn't work for you, think of learning a language. It might be a scientific one, or a language of bureaucracy, but you need to try to become fluent. Why? This is for two key reasons. How can you possibly know if something is amiss unless you understand what things are supposed to look like when everything is going right? And why should sources with a lot to lose risk giving you information if they have to educate you from the ground up?

Take an example from Dan Diamond's terrific reporting at Politico. In September 2020, he broke a big story: Inside the federal government's Health and Human Services (HHS) agency, a Trump political appointee, Michael Caputo, had sought to alter scientific guidance put out by the Centers for Disease Control and Prevention: the *Morbidity and Mortality Weekly Report* (*MMWR*), which had long been free of political interference.

How did Diamond know to look for possible changes to the *MMWR*? He didn't, specifically. Instead, he'd been on high alert when 5 months earlier, Caputo, a political operative, had been installed inside HHS. "In normal times that's an eye opener," Diamond says. But, he also says that "in a pandemic to put a political communications expert in the middle of a crisis where his impulse is all political, I knew from that first day" how unusual it was. In other words, Diamond knew what normal looked like. And so to him, Caputo's installation was a red-flag event, a divergent trail that he followed.

The lesson is to spend time learning the landscape. When I'm stuck, or no one's returning my calls, I return to remedial landscape studies. I've watched promotional videos of generic drug manufacturing plants on YouTube and cracked open the Food and Drug Administration Title 21 *Code of Federal Regulations*. Nothing you learn will be wasted in the end.

Tracking the Information Flow

When the pandemic struck, you'd think I'd be well prepared to cover it, having reported around medicine, pharmaceuticals, and the federal health bureaucracy for a long while. But as I jumped into the biggest disease story of a century, I was thoroughly lost. *PPE*—What was that? *The wild-type virus*—Huh? What, exactly, does an epidemiologist do?

I was in a foreign world without a clue. How was I going to cultivate high-level government sources who were not supposed to speak with me, and almost certainly wouldn't, if they thought (a) I had no information and (b) they were going to have to educate me.

This is the part of the reporting that I think of as crawling across the floor in a house that's on fire. There's smoke everywhere. You can't see more than an inch ahead. Instead of panicking, stay low. Go back to the absolute basics. What do you know, and how can you leverage it? In my case, I took out a very old Rolodex with yellowing business cards and started tracking down sources from long ago.

In this painstaking manner, I began to hunt for ambassadors—people who could introduce me, and vouch for me, to crucial experts. It was a moment when everyone wanted to help, so some of my old acquaintances made a real project of it. They introduced me to people who, over time, became indispensable sources.

I also threw myself into learning everything I could about this new world. I began hunting for organizational charts of federal government agencies, each of which had some role to play in the response to the pandemic. What was each one responsible for? What were all those different divisions of HHS? Who were the past employees who might have mobile numbers for the current ones?

You can't climb Mount Everest by trying to lasso the summit. As hundreds of news outlets trained their sights on the top few sources like Dr. Anthony Fauci, I was busy studying the lower base camps and the pathways between them. In short, there are numerous different ways to ascend, each with hundreds of different footholds. I was studying the organization charts to figure out where the information was and who had it.

That effort left me sleepless at night, as I tried to map out the flow of information and documents. I envisioned a wheel with spokes. In the center was the information I wanted. Each spoke was the agency or person who might have it.

Finding and Keeping Sources

So, you're no longer totally lost, and you've educated yourself. You know the landscape, you can speak the language. But how can you find and recruit sources to lead you to relevant stories?

I always look for what I think of as spirit guides: topic area experts who are natural teachers and motivated to help. During my reporting for *Bottle of Lies*, I leaned heavily on the input of several corporate whistleblowers. Of course, I still had to learn how to analyze things myself, but when someone with real knowledge says, "Hey, look over here," that can be critically important.

Once you've figured out who has actual information, you might assume: "They'll never talk to me." But you won't know unless you try.

First, you'll want to try approaching them in their favorite medium. If someone has a dormant Facebook page, a message there is unlikely to work. But if they're posting night and day on LinkedIn, or you learn that they live on Signal, that's the best approach. Whenever possible, I try to get input on their communication habits from their colleagues. I was recently told that a potential source would likely not respond on Signal unless I set the "disappearing messages" bar to 12 hours. Boy, was that good information. Once he did respond, he set his messages to disappear within 5 seconds (I almost crashed my car trying to read one of his messages before it vanished).

But let's say you make the approach and get no response: don't give up. Another method is to try acting as if you have a relationship—until you actually do.

To one prospective source, I continued sending messages through LinkedIn, maybe 10 in total and got no response. My messages were always friendly and even keeled. Then one day, he responded, and his information enabled me to blow a story wide open. But here was the key: I didn't just ask for information, I offered it as well. The prospective source finally answered me because he felt strongly about some of the information I shared and knew I was on the right track. In short, I'd demonstrated my knowledge.

You also need to understand the role of loyalty tests in cultivating sources. There had been a source I was angling to speak with during my reporting for *Bottle of Lies*. Through an intermediary, I got a message back: He's willing to speak with you, but he'll only do it in person, in Beijing. Travel from New York to Beijing for a single conversation with a source I'd never met? That's crazy! But doing it was the best way to demonstrate my commitment to the story. I got on a plane to Beijing, a decision that dramatically improved my book.

Sometimes the challenges are explicit. One time, I'd been pursuing a prospective source inside a federal government agency. He responded by email: "I'm still not convinced that this story should be told, and I don't understand the purpose. I'm also not convinced that you are the person who is ready to tell this story. Not that it should be someone else, but you haven't demonstrated your knowledge . . . to me."

I recognized this not as a rejection, but as a challenge. Within 3 months, he handed me a thumb drive with 20,000 documents on it. He had copied his files so thoroughly that the documents included his divorce proceedings (which I deleted). That told me that I'd earned his trust.

Once you've established a relationship, you must be worthy of it. Don't just ping your sources when you need something. Check up on them. Ask how they are. Any number of times, I have recommended against publishing something that a source was willing to have me print because I thought it risked exposing them. My job is not only to publish information, but also to look around corners for my sources, candidly assess their risk of exposure, and protect them at all costs.

Getting Documents

Have you ever tried putting up a tent in a Category 5 hurricane? Probably not, but one thing is guaranteed. You'll need tent stakes, and each one makes your tent a little more secure.

Documents are those tent stakes, and to be meaningful, they don't need to just be an internal memo where the company chief executive officer admits to fraud. Calendar entries, invitations, event programs, fliers, and attendee lists—anything that helps to document people's activities and actions—help to strengthen your story.

But documents can also accelerate your reporting in less obvious ways. They can serve to cement relationships with sources. Once someone gives you a document—even if it's a public record you could have gotten off a website—that gives you an opening to return to ask follow-up questions. Documents can also act as a chisel. They increase the likelihood of a return phone call from a potential new source. Imagine the difference between the messages you can leave for someone: (a) "I am trying to find out if you were at the Century Club on August 9." or (b) "Calendar entries I obtained indicate you were at the Century Club on August 9. Would you have a minute to speak with me about that?"

In some cases, documents can serve as doorways into a completely hidden world. In reporting on Covid-19, a trusted federal government source passed along something that struck him as odd: a one-page invoice from a United Arab Emirates company for 3.5 million Covid-19 diagnostic tests for $52.5 million. The most arresting part of the document was the entity that had ordered the tests. The client name simply had two initials: W. H.—the White House.

The document was a mystery. The White House couldn't procure anything. By law, all federal government contracting took place under a rigorous system, with orders only approved by a duly authorized contracting officer.

The clues embedded in that invoice ultimately led me to uncover a far bigger story. A White House task force overseen by President Trump's son-in-law and special advisor, Jared Kushner, had drafted and then buried a comprehensive national Covid-19 testing plan, in part because of a political calculation that the pandemic was only affecting blue or Democratic states. Over weeks of back and forth with a new source, I obtained a copy of that plan. The resulting story was the most widely read in Conde Nast's history.

The Takeaway

The worlds of science and medicine—awash with money and life-and-death stakes—offer the perfect terrain for investigative journalists. There are whodunits to unfurl, regulatory breakdowns to reconstruct, and a surprising number of whistleblowers who feel compelled by conscience to speak out or leak documents. As you become more expert, you will be able to find and break stories that few others will.

It's not for the faint of heart. Sometimes, the stories are so complex, and the consequences of getting it wrong so serious, that you might feel like you're doing open heart surgery yourself. But the most rewarding part of the beat is that your stories actually have the potential to save lives or expose what took them in the first place.

16

Accessing Public Records

Michael Morisy

Michael Morisy is the cofounder and chief executive of the MuckRock Foundation, a nonprofit that helps journalists, researchers, and the public file, track, and share public records requests. In addition to MuckRock.com, the organization runs the DocumentCloud, oTranscribe, and FOIA Machine services, used by thousands of newsrooms around the world.

It was a basic, if morbid, request: After a dolphin died in her home state of New Jersey, a MuckRock intern filed a request for the autopsy. If the state had simply provided the records, it likely wouldn't have even merited a write-up. But as government agencies often do, the New Jersey Department of Agriculture stonewalled and denied the request, citing the need to protect the dolphin's medical privacy rights, thus providing a series of illustrated lessons for Freedom of Information Act (FOIA) aspirants.

But before we get to those lessons, here is a brief primer. Passed in 1967, the FOIA grants access to the records of most federal agencies unless they can cite a valid exemption. Congress, the courts, and the White House itself are all generally exempt with a few minor caveats.

In the wake of the law's passage and post-Watergate reforms that strengthened it further, every state eventually passed broadly similar FOIA laws, which have their own quirks: Most let you file requests with legislative offices or the governor, for example, but some don't. In Louisiana, a requester must be at least 18 years of age. Some state laws make contractors or nonprofits subject to FOIA laws if they're doing work on behalf of the government, while others make no such stipulation.

While freedom of information took almost 200 years to be written into America's law, there is something almost uniquely democratic about it. Compared to so many other rules and regulations, it puts power in the hands of the people. You get to tell the government what to do, and unlike when you're shouting a question in a scrum or submitting a request for comment, officials are legally obligated to respond.

Michael Morisy, *Accessing Public Records* In: *A Tactical Guide to Science Journalism*. Edited by: Deborah Blum, Ashley Smart, and Tom Zeller Jr., Oxford University Press. © Oxford University Press 2022. DOI: 10.1093/oso/9780197551509.003.0017

Even the Founding Fathers recognized the fundamental importance of access, complaining in the Declaration of Independence that King George "called together legislative bodies at places unusual, uncomfortable, and distant from the depository of their public Records, for the sole purpose of fatiguing them into compliance with his measures."

Thanks to international trade agreements that often stipulate public records laws, almost every country now has some version of them on the books, but their effectiveness and independence vary widely. Authoritarian countries often apply sweeping and vague security exemptions that do more to cover up embarrassment than protect national security. In India, people have even been killed after filing basic requests for information on property records, a chilling response regarding what should be a basic civil right. In America, some states have begun clamping down on access, requiring requesters to show ID, for example. And even at the federal level, agencies don't always comply with the law, meaning the fight for access often requires vigilance and persistence. As a requester, you play an important role in keeping the process healthy.

In principle, transparency laws should offer everyone equal access—it's our shared government, after all. But journalists still have a special place in the process, generally acting as the public's proxy in transparency fights. That can mean lower fee categories and waivers, as well as expedited processing in rare cases. When you get discouraged, keep in mind that improper public records denials—common as they are—are not just a frustrating nuisance, but an affront to an informed democracy.

The challenge is that while you can ask your government about anything you want, freedom of information laws only pertain to documents or data. As a result, you'll need to develop some sense of the concrete material you're after and know how to tailor your request so that it produces results. In the end, that's precisely what broke open the case of the New Jersey dolphin—but we'll get to that.

For now, just know that public records can be an endless source of scoops, whether you're new to the beat or a senior correspondent.

The Dirt Is in the Details

There's no record too small to request. Governments run on bureaucracy, and bureaucracy runs on documentation. And, unless the government agency can provide you with a reason not to hand it over, you have a right to that documentation. Let your curiosity guide you to request reports off-handedly mentioned in a press conference, studies cited in an obscure footnote, and, yes, the postmortem for a bottlenose dolphin.

In fact, being able to pick up on the right details is what makes both great writing and great FOIA requesting. A common mistake is treating records requests like Google, where you can slap "any and all documents about" around a topic you're interested in and get good results. Sometimes, you'll get lucky. Usually, though, the agency will chuck out your request as too broad or too burdensome or, worse, banish it to the complex processing queue.

To avoid that, take a minute and bring your creativity and research skills to the process.

Try to imagine being a middle manager in a beige government office. Visualize what data are being gathered and archived by your department in its daily business. If you were overseeing some program you're reporting on, imagine what kinds of presentations, requisition forms, sign-offs, budgets, proposals, and other errata would be involved.

Consider, for example, a 2012 investigation into speeding cops by Sally Kestin and John Maines for the *Sun Sentinel* in Fort Lauderdale, Florida. It started with a common observation: Police cars could often be seen cruising the turnpike at head-spinning speeds, even for Florida, with no lights flashing.

Since Florida's roads are tolled, and cops receive special transponders letting them pass through the tolls free, the reporters were able to request, under Florida's expansive Sunshine Law, the exact time and place where officers entered and exited the Turnpike. With that data and a little math to calculate how long it took them to get from Point A (the entrance to the turnpike) to Point B (the exit), they were able to show just how flagrant unnecessary speeding had become. They then requested work schedules and showed that officers would routinely drive 25, 40, and even 60 miles per hour over the limit to and from work.

Each state has what's called a records retention schedule, detailing the types of documents various agencies are required to keep and how long they need to keep them. There are also agency records system lists, which cover most of the databases you can request. Few journalists spend the time to read them, but those who do get a wealth of great request ideas. For example, you can ask the General Services Administration for a spreadsheet of virtually every car that the federal government owns. In many communities, you can request a copy of dog registration data to see the top breeds or the database of city-owned trees (is the town adhering to its environmental pledge?).

You can also file meta-requests, such as for copies of any blank forms that an agency routinely gets printed up or even FOIA logs themselves. These indexes of other people's requests can be extremely useful for inspiration. In fact, if you see something that might be potentially interesting, the FOIA office will usually have a copy of those preprocessed documents handy and can forward them to you very quickly.

The FOIA logs can also be great for source development. Most records requesters aren't actually journalists. Looking up other requesters' names can often connect you with colorful characters, ranging from jilted contractors with an axe to grind to attorneys and activists digging into dark corners of agencies you didn't know existed.

Speaking of sourcing, there is one routine request you might file with any government agency you plan to cover: copies of recent employment offers, resignation letters, and dismissal notices. While the letters themselves are usually not very revealing, they give you a peek at the revolving door and potentially an introduction to recently departed staffers who may have a story to tell.

Don't Take Their Word for It

There are obvious power imbalances in the world of public records. The government, after all, is in actual possession of the documents you're after, and you might only have a vague idea that they might or at least should exist, let alone which specific filing cabinet they're in. To make matters worse, there's an experience gap: An agency receives hundreds, thousands, even tens of thousands of requests annually, while you might be filing your first.

Don't forget, however, that you have your own points of leverage. First, records laws generally state that the default is disclosure unless the government can cite a valid exemption: They have to prove why they can keep the documents from you, instead of you having to justify their release. Second, the public records process is built on a paper (or at least electronic) trail.

In the case of the poor beached dolphin, New Jersey cited an exemption—medical privacy—that was not actually valid in this case, as the deceased have limited privacy rights and dolphins have none.

Agencies mangle or invent exemptions surprisingly often. As a result, it pays to spend a little time comparing the exemption the agency cites (and they did cite an exemption, right?) with what the law actually says. If they cite a legal code, go ahead and Google it, or check out the excellent and accessible Open Government Guide of the Reporters Committee for Freedom of the Press (RCFP), which details how exemptions are properly applied at the state level and also provides plenty of legal citations that narrow seemingly expansive roadblocks.

For example, agencies love saying that documents are being held back because they're "deliberative" or "predecisional," an exemption designed to allow brainstorming and candid discussion before decisions are made. But deliberative process does not apply to the basic facts an agency used to guide its discussion, so if the agency compiled internal data to guide its thinking, they must still release that—even if it can withhold other parts of a document.

Another common cause for rejection is that the agency says it can't find the documents. This is where your attention to detail pays off. When drafting your request, look for citations or other evidence that the document exists and make sure to make it part of your request. A number of studies have found that FOIA requests that include links to related materials get handled more quickly and completely than requests without. The reason is that the additional context helps the records officer more quickly and efficiently understand where to go internally and makes it harder for the agency to deny the documents' existence outright.

Finally, agencies will often assess large fees for records requests, or at least fees that feel large to your typical journalist's budget. At the federal level, make sure they've properly classified you as a media requester, which restricts them to charge only for direct material duplication costs (i.e., paper and ink) after the first 100 pages, which are free. If they miss their deadline without being able to prove your request was particularly complex, they can't charge any fees.

State laws are generally not as forgiving, but check to see what rules apply (more on that further in this chapter). Agencies tend to have a very different perspective on what reasonable fees are, so usually the best response is to ask questions: Is there a portion of your request that's leading to most of the fees, and do they have suggested modifications? Can they give you the first 10 pages of responsive materials so you can see if it's actually what you want? What if you come in and view the documents in person?

Most states offer a right to inspection, and if there's one thing agencies find more annoying than processing freedom of information requests, it's having a reporter hanging around the office for hours snapping photos of documents.

Fight Back When Necessary

You've done your homework, you've filed your request, and bam—your quest for transparency slams into bureaucratic indifference or hostility. This is when you need to lift your game a notch.

First, figure out what appeals processes might apply. This is where state and federal laws sharply divide. At the federal level, you're entitled to an administrative appeal as well as independent mediation, with a window of about 90 days after a request is rejected. In their rejection letter, agencies will even include instructions on appealing so you know exactly what to do and when to do it.

State rules vary widely. In Connecticut, you can file an appeal with the independent Freedom of Information Council (what a nice ring). In Florida, you'll have to jump straight to litigation, while in Texas an agency actually has to run rejections by the attorney general before denying you access.

Don't be intimidated by the appeals process. MuckRock has a ton of free guides to each state's laws, as does the RCFP. Also check out your local chapter of the National Freedom of Information Coalition (NFOIC), which will likely run a hotline with a lawyer happy to help (it's amazing how often legal letterhead will have an agency change its tune).

In most cases, though, you don't need a lawyer to file an appeal. Just explain your reasoning as clearly as you can, whether you think a law was misinterpreted or the officials didn't look in the right places, and fire away. The most it can cost is a stamp, and by some estimates as many as one third of appeals result in the release of more records.

If you do want a lawyer, there's an increasing number of resources. Along with the NFOIC and RCFP, which both often have lists of pro bono or low-cost transparency attorneys, check out the Free Expression Legal Network, a national coalition of law schools that connect top-notch students under very accomplished supervision with thorny cases.

But as a reporter, don't forget your most powerful tool: your soapbox. If you make a reasonable request for a document and the agency can't give you a reasonable response, write about it and why that information matters. Tweet out ridiculous redactions or justifications. Raise the alarm and see if you can get other reporters to join the cause.

This is what ultimately helped dislodge the story behind our dolphin's fate and the agency's peculiar protectiveness around its privacy. The absurdity of the argument turned what would have likely been a nonstory into a national one, with outlets across the country covering the rejection, and that's when the real scandal came out.

After ceaseless mocking inquiries, someone finally fessed up. The real reason for the rejection was that the animal rescue nonprofit that requested the autopsy from the state had been accused by an environmental group of being too aggressive about euthanizing animals. The organization pressured the state to withhold the information, so the state scrambled for some excuse, however implausible, to hide the documents.

Caught between the truth and unexpected public pressure, the agency ultimately released the documents, showing the cause of death as morbillivirus— what would have been a much less interesting story if they had simply released the documents in the first place.

The Takeaway

There are countless documents with interesting stories to tell squirreled away in filing cabinets, government warehouses, and agency hard drives all around the world, just waiting for a requester clever, patient, and curious enough to ask.

Set aside time each week to let your mind wander around what interesting databases might exist, what reports might be getting filed, which hotline complaints are logged, and ask for them.

They key is to turn each rejection into a lesson and to let each success inspire follow-up requests. When officials promise to complete a report in 6 months, mark your calendar. When filling out a mindless bureaucratic form, imagine the database or filing cabinet filled with stories just waiting to be told. See government hotlines, surveys, and even surcharges and fees as opportunities to dig a little deeper, as there is always a record and potential story on the other side.

Even the simplest request can take you on strange, unexpected journeys, but only if you file it.

17

The Art of the Interview

Pallab Ghosh

Pallab Ghosh is a science correspondent with BBC News. He works across television, radio, online, and digital platforms. He has won numerous awards, including Press Gazette Science and Technology Journalist of the Year, the United Kingdom's equivalent of the Pulitzer Prize, for investigative reporting, and is a former BT Technology Journalist of the Year. Pallab has a strong interest in the societal implications of developments in science and technology and aims to involve people in shaping policies in controversial areas such as genetically modified crops, cloning, and artificial intelligence. He is also honorary president of the Association of British Science Writers and a founding board member and then president of the World Federation of Science Journalists. In these roles, Pallab has promoted critical coverage of research issues. He believes that for science to truly benefit society it must be tested by penetrating science journalism of the highest professional standards.

Journalists have many privileges. For me, the greatest one is the opportunity to engage with people about their ideas, passions, and dreams. I have had the opportunity to interview a number of notable people, including Professor Stephen Hawking, whose devilish sense of humor constantly played across his eyes; Vint Cerf, one of the "fathers of the Internet," who gave a small bow and humbly replied, "One is glad to be of service," when I thanked him for the web; and Jim Lovell, who radiated both kindness and steeliness as I heard firsthand of his experience bringing the crew of Apollo 13 back home safely.

But my encounters with lesser known subjects have been equally rewarding, like when a theoretical physicist told me that results from Fermilab about a possible "fifth force" of nature was what he had waited for all his career, and that he was so excited that he was unable to sleep. Or the modest astronomer, who when asked what he would do to celebrate the confirmation of groundbreaking ideas he developed long ago as part of his PhD thesis on the evolution of the early universe, told me that he would have pizza with his family.

Pallab Ghosh, *The Art of the Interview* In: *A Tactical Guide to Science Journalism.* Edited by: Deborah Blum, Ashley Smart, and Tom Zeller Jr., Oxford University Press. © Oxford University Press 2022. DOI: 10.1093/oso/9780197551509.003.0018

Such revelatory moments, whether from the legends of science or work-aday researchers, can enrich and deepen readers' understanding of a story. But they don't come about by chance: They are the fruits of careful preparation and a strategy to develop a relationship with your interviewee.

Your ultimate goal is to get the best out of your subjects, so they can say what they really believe and feel, rather than what they think they are supposed to say. To do that, you have to ask the right questions. It is as simple as that. Figuring out what those questions should be is where the art of the interview comes in.

Preparation

You usually have a set amount of time with your interviewee. Don't waste a second of it by asking questions that could have been answered by doing a few simple internet searches ahead of time.

This process can reveal that the story you are pursuing has already been done, in which case you either drop the interview or think about how you can advance or deepen it. If, on the other hand, your search reveals that your story really is new, you can focus on the novel aspects and how they sit in the broader context of the field.

Then there is the crucial step of formulating your questions. I find it helpful to write a short summary of the story so far. This establishes the gaps in my knowledge and helps me determine some lines of enquiry. You may not know the details of your story at this stage, but you should anticipate its overall shape and structure and where your interviewee is likely to fit into it. This will give you focus and save time.

You should also research alternative perspectives and even criticisms of your interviewee's work and put those points to him or her. Remember: You are a journalist, not a public relations (PR) conduit, and it is your job to challenge your interviewee where appropriate (more on that further in the chapter). Their responses should make your story stand out, rather than weaken its content.

The Relationship

Any salesperson will tell you that before talking to a potential customer about a purchase, you have to develop a relationship. It is the same with journalism.

You are asking your interviewee for more than just a transaction. The reporter is often asking for trust and empathy. This starts before you ask your first question.

You want to create a good impression and to make your interviewee feel comfortable. So, be presentable and don't be late. Along with your wits, courtesy is the journalist's most effective tool.

What you wear matters. Scientists and science reporters often tend to be fairly casual, but sometimes there will be interviewees who prefer greater formality. If you are not sure, err on the side of being slightly conservative in your appearance.

Most important, though, is what television interviewers call handling the subject. Typically, this is the interaction you have in order to settle them down while the camera crews are setting up with lights and framing shots. But it is something I also do for print, online, and radio interviews because it always gets better results.

The trick is to get your interviewees to forget they are giving an interview and encourage them to feel they are just having a relaxed chat. Start off by having a casual conversation before the interview begins. Take an interest in the person and listen. After a few minutes, suggest, almost in passing, that you might as well begin the interview and then continue in the same conversational manner.

In those first few moments, try to assess your interviewee to determine how you conduct the interview. Is she shy and requires a little coaxing? Or suspicious and needs to be won over? Or perhaps she is overtalkative and needs to be kept on track.

Handling your interviewee is essentially a getting-to-know-you process—a way to show courtesy and respect. That goodwill will usually be returned and thus enhance your interview.

Interviewees will also have prepared for their interview. Some will have overprepared and be ready to deliver a scientific seminar. It will be useful, but unlikely to give you the information you need—that is, why the story is important, why the researcher is excited about it, and how he or she feels about it.

I try to establish this key information in an informal conversation ahead of the interview, often in the handling process. Once I have discovered these nuggets, it is easy enough to gently steer the interviewees away from their presentations and on to the things that really get them excited.

In cases where I suspect my subjects might be stiff, I begin recording them while having the informal conversation before the real interview starts. Sure enough, their eyes will light up, they'll tell you funny anecdotes and often give you a vivid take on why the thing they are talking about is so important.

We then proceed to the actual interview, which is usually more reserved. I always ask afterward, of course, if I can use sections from the preinterview, and almost always the interviewees are delighted that I recorded the parts in which they felt they expressed themselves more naturally.

The Interview

Remember the first interview you ever did, maybe when you were playing reporter as a kid? You may have written a list of questions and asked them one after another, regardless of the answer:

Reporter: What was the key breakthrough that led to you winning the Nobel Prize?

Scientist: I was getting nowhere with my experiments, so I made up all my data.

Reporter: What would you say drove you on?

Scientist: The desire to build a secret arsenal of weapons of mass destruction.

Reporter: Which scientists have inspired you the most?

Scientist: . . .

The right question is often not the next one on your list. Sure, you have to have some in mind, but listen to the answers and ask pertinent follow-ups. This may take you in unexpected directions.

There have been occasions when I've interviewed a scientist and something sensational emerges, possibly the work of a colleague or something heard on the grapevine. Fabulous though that is, it is more usual for conversations to take a turn when the researcher highlights an aspect that is more interesting than you had appreciated. So, don't stick to your script.

You should have a starting point and a route sketched out, but think of your interview as a leisurely stroll with a friend, where you take your time at points of interest, skip past the dull stuff, and take detours if another route seems more appealing.

Your questions should ideally be reshaped by each answer, usually subtly, sometimes radically, so that your interview takes you further along the journey that you and your subject are taking together.

Context is also important in determining your questions. Some interviews may simply comprise a couple of barked questions at a news conference, while others are analytical dialogues of gentle exploration. Then there are the different platforms to consider. Broadcast interviews need to be about performance as well as content, while those for social media need to be sharper and briefer.

Interviews are also about your perception of the interviewee. This will be reflected in what they say in response to your questions, but there are plenty of physical cues to look out for that will be just as illuminating. Do they fidget? Do they appear sad or awkward when responding to a question? If so, how do their emotions manifest themselves? Watch out for half-hidden signs of tension in the verbal and body language of interviewees. Following up on small indicators like these can sometimes yield interesting lines of questioning. Record your observations and add them to your story.

The most important factor in a successful interview is that there is no such thing as a stupid question. When you are starting out as a reporter, you might feel embarrassed to ask researchers to explain their work at a nontechnical level and, crucially, to stop them and ask them to go over it again in even simpler terms if you don't fully understand. If you can't follow what your interviewee is saying, you are not going to be able to explain it to your audience. And chances are your editor will spot this and ask you to go back to the subject for further explanation. Only stupid people don't ask stupid questions.

Finally, having completed your interview, don't forget to get the basic facts, such as the person's age, key dates, spellings, and so forth. It is a tedious but essential part of the process.

The Adversarial Interview

A former BBC presenter, Jeremy Paxman, was infamous for mercilessly tearing apart politicians in live TV interviews. The ratings were through the roof for the show, with people tuning in for what was essentially a gladiatorial spectacle. When asked about his technique, Paxman responded by saying that during his interview he asks himself, "Why is this lying bastard lying to me?"

Scientists are not politicians, however, and are usually not deserving of such treatment. Occasionally, though, you may be faced with an interviewee who does need to be grilled about an issue where they might be at fault or at least be obliged to explain their actions.

Faced with an individual or organization accused of wrongdoing, your starting point should always be to get their side of the story. If you have a reputation for balanced reporting, an individual under fire may well grant you an interview because they will be assured of fair treatment. In these circumstances, the approach is to simply put the points of contention to your interviewee and challenge where appropriate.

There will be situations where you won't get a straight answer. So, don't expect one. If your interviewee initiated the interview, he or she will usually have been briefed in advance to stick to certain answers. Your job in those situations is to expose whatever wrongdoing there might be through your questions. State the charges clearly and firmly, as if you were a prosecuting attorney. The preprepared responses will typically seem weak and inappropriate, and the audience will be able to draw its own conclusions about culpability.

In some situations, it is legitimate to rattle an interviewee with an assertive line of questioning. If you are not used to doing this, role-play the interrogation with a colleague beforehand. This can sometimes sufficiently annoy your subjects and cause them to stray from their PR-prepared script and launch a tirade. Or it can go the other way, putting an interviewee at a loss for words. Even in print and online stories, a description of the interviewee's unease will further convey doubts over credibility.

The Big Interview

The opportunity to interview a prominent person is one of the high points of any journalist's career. I've been fortunate to do this with a number of my scientific heroes, and I'm one of the very few journalists to have interviewed Neil Armstrong. I find it helpful to speak to a variety of colleagues and editors as part of my preparation, but I always ask myself, "What do I really want to know about this person?"

It may well be a once-in-a-lifetime opportunity, and you will kick yourself forever if you miss the opportunity to ask some of the burning questions on your mind. Many prominent people will have given many interviews before and will be prepared with standard answers to obvious questions. What will be unique is their interaction with you. Bring your own personality: Showing your passion and enthusiasm in your questions will veer your interviewee away from pat answers and usually bring the best out of them.

With Armstrong, there were many things I wanted to ask him. But I only had time for a few brief questions. The one I most wanted to ask was: "When you first set foot on the moon, we thought that this was just the beginning of an exploration of other worlds, the very stars seemed within our grasp. That dream has now gone. What ever happened to the Armstrong dream?"

He gazed toward me, and his kindly presence filled my vision. He said: "The dream remains! The reality has faded a bit, but it will come back, in time."

And with that reassurance, he set my world straight again. Once again, I could dream about the possibility of exploring the stars!

It was a deeply personal moment, but one I was able to share with a wide audience, and I hope it brought something to the story that no one else could.

The Takeaway

Having a conversation is as natural as breathing, but there is far more to it than that for a successful interview. It is much more than a process of simply acquiring information. Each interview is unique, sometimes intimate and always a privilege. Get it right and you'll get a better story, even if it's often not the one you were expecting.

18

Cybersecurity and Protecting
Your Sources

Andrada Fiscutean

Andrada Fiscutean is a science and technology reporter based in Bucharest, Romania. She has written about Eastern European hackers, journalists attacked with malware, and North Korean scientists. Her work has been featured in Nature, Ars Technica, Wired, Vice Motherboard, *and* ZDNet. *She's also editor in chief of ProFM radio in Bucharest, where she assembled a team of journalists who cover local news. In 2017, she won Best Feature Story at SuperScrieri, the highest award in Romanian journalism. Passionate about the history of technology, Fiscutean owns several home computers made in Eastern Europe during the 1980s.*

While reporters covering political dissidents, international conflict, and global spy rings might fully expect sophisticated attacks on their private communications, science journalists don't generally think of themselves as potential targets for information-seeking hackers—and in many cases, they are probably right. But it's also worth understanding that science-based beats do sometimes delve into topics and subject areas—from nuclear weaponry to computer espionage and even corruption in science itself—that *can* place science journalists in the crosshairs of nefarious actors keen to access and exploit their computers and cell phones.

Communications with anonymous scientific whistleblowers, for example, are not uncommon for some science reporters—and the handling and storage of sensitive digital documents are also common. And if you think you are already being careful enough, it might be worth thinking again.

Indeed, there's an adage in the world of cybersecurity that there are two types of people: those who have been hacked and know it and those who have been hacked but don't know it. When it comes to journalists, hacking can often be invisible. Governments and corporations, for instance, are less interested

Andrada Fiscutean, *Cybersecurity and Protecting Your Sources* In: *A Tactical Guide to Science Journalism.* Edited by: Deborah Blum, Ashley Smart, and Tom Zeller Jr., Oxford University Press. © Oxford University Press 2022. DOI: 10.1093/oso/9780197551509.003.0019 in stealing our money than curious about the stories we are working on or the sources we talk to.

Journalists can be powerful, but they are also vulnerable. And hacking them is "shockingly easy," says freelance technology reporter Eva Wolfangel.

Many of the attacks are unsuccessful against vigilant journalists, but an attacker only needs to deceive us once to take over our computers or phones. It sometimes feels like a battle we're bound to lose, given the resources and persistence that are often involved. But in fact, the skill that can support us is embedded in our journalistic standards. We just need to be skeptical and double-check all our emails, even those coming from our moms.

Simple and Secure

I've been covering cybersecurity for a decade now, and I often remember Yevgeny Zamyatin's dystopian novel *We*, which portrays a society in which people live in apartments made almost entirely of Panopticon-like glass structures that allow for constant surveillance. When it comes to our digital lives, we might not be that far away from that future.

Yet, there's not just one "Big Brother" watching us. We're living in a "Some Brother" society, where someone is always watching, according to Finnish futurologist Mika Mannermaa: "Some Brother is controlling, knowing and never forgetting."

This environment leaves us little room for mistakes, but there are a couple of things we can do to improve our security. Antivirus software, password managers, and multifactor authentication are some of the basic tools we can add to our arsenal to relieve technology-related anxiety. Yet staying safe is not only about using a set of tools, but also reliance on adopting a security mindset that leads you to question everything, from the apps we download to the emails we receive. Innocent-looking messages not only can include personal details such as the schools our children go to or our previous stories, but also can imitate the newsletters we subscribe to. A hacker can do virtually anything to trick us into opening attachments or clicking on links.

So avoid clicking on unfamiliar links and opening attachments, especially those coming from strangers. Wolfangel tells her sources to send information in clear text, in the body of the email, as most cyberespionage and cybercriminal campaigns still rely on malicious emails.

When there's no clear workaround and she must open a suspicious file, Wolfangel first uploads it to platforms such as VirusTotal.com, which analyzes the file with a few dozen antivirus software products that check for malware and issues a report within a couple of minutes.

Sometimes, these basic rules can make or break a story, as happened to me a few years ago, when I approached the well-known cryptographer Jean-Jacques Quisquater, who wanted to test me before agreeing to be interviewed. After I sent him the initial interview request, I got a fabricated email that looked just like a LinkedIn invitation coming from him. Had I clicked the "Accept" button, I would have lost him as a source. Luckily, I knew that a fake LinkedIn invitation sent by email was exactly how he himself got hacked.

Often, staying safe isn't about doing more, but doing less. This includes not clicking on strange links, of course, but also installing fewer apps on our phones, or even discarding smart devices—whether TVs, smart speakers, or baby monitors—that constantly listen to our conversations. Security researcher Mikko Hyppönen likes to say that whenever we hear the word "smart" in relation to a device, we should replace it with "hackable." Our laptops and phones also include microphones, so if a resourceful organization hacks them with advanced tools such as Pegasus, they could remotely listen to our conversations. This is why it might be good not to read our investigative stories aloud when writing them.

In a world where everything can be attacked at any time, protecting ourselves can feel daunting. Yet as soon as we start doing it, it gets easier. Little by little, we internalize the additional steps that keep us safe, better protecting ourselves and our stories.

Hidden Writing

Humans have always looked for ways to hide messages. Julius Caesar, for instance, invented a cipher that takes the Latin alphabet and shifts characters by three places, so A becomes D, and B becomes E, and so on. He used this simple method to encrypt his personal correspondence, and some historians believe it was quite effective for its time.

Probably the most famous encryption tool is the Enigma machine, developed by German engineer Arthur Scherbius. Many historians say that cracking its code helped the allies to substantially shorten World War II.

Today, we all have access to encryption tools that are even more advanced than the Enigma, and we can use them to secure our chat messages and emails. For many journalists, doing this is vital, and reporter Charles Piller of *Science* magazine uses them at all times. He says a journalist can never be too careful, and knowing how these tools work also helps him gain the trust of those who want to share critical information. "It helps to meet a source at his or her level of paranoia," he says.

Piller encrypts not only his conversations, but also his hard drives. This way, if they get stolen, the information would be virtually useless. Encrypted hard drives are his go-to method of storing information because he doesn't like to upload files to the cloud. "I don't trust online storage services with sensitive documents, or documents that I wouldn't want in the public domain," he says.

Some journalists, including Piller, take it to the next level and use a so-called air-gapped computer to research and write their stories. This kind of self-contained machine, which is unconnected to the internet, is even more secure since it cannot upload files to a potentially shady server.

But journalists not only store, process, and create documents but also sometimes have to destroy them securely, and normal computers are awfully bad at helping achieve that. When we delete a file, even from the trash, that file remains on the hard drive. The only way to get rid of it completely is by overwriting the disk sectors it occupies, 20 or 30 times, using special software to make sure it is beyond recovery. Of course, there are also drilling machines that can create holes in the platters of a hard drive to make the information stored there unsalvageable.

Unfortunately, many people who want to give us tips or documents don't use encryption. This puts everyone at risk, says Piller. He maintains that it's our duty as journalists to educate sources and editors about communicating securely and handling information. As a result, whenever he gets a message from a source, he tells them right away to download encrypted messaging apps such as Signal, Threema, Ricochet, or Confide, and continue the online call or conversation there, in an encrypted form.

"I balance my need to get information quickly with a sense of ethical responsibility to help naïve sources understand how to protect themselves," Piller says.

As security researchers like to say, "Dance like no one's watching, encrypt like everyone is."

Travel Securely

Stories can take us on journeys across the world, but there's a rule for journalists: Don't bring anything you don't need. Reporting trips can be unpredictable, especially when covering countries that have thorough border control or surveillance.

Investigative journalist Katherine Eban recalls going to Beijing to interview sources for her book, *Bottle of Lies: The Inside Story of the Generic Drug Boom*, in which she reveals how some drug manufacturers in China and India falsify data and cut corners. Though she is an advocate for strong digital security, she nonetheless got hacked.

She remembers being in a hotel lobby, waiting for a source to arrive, when the home screen of her phone changed. It showed a photo featuring a Chinese man holding up an English language newspaper. The phone was the only personal electronic device she had brought with her during that trip; she left everything else at home, including her laptop. She thought buying a SIM card on the Chinese gray market would be enough to keep her safe. Apparently, her attacker had enough resources to try harder.

When I take reporting trips to meet technology professionals in Eastern Europe—particularly in Russia and Ukraine—I try to leave all my equipment at home and carry only a burner phone and cheap laptop, which I can discard after a single use. For these, I have a separate email account, never logging on to websites with my usual credentials, and stay away from Facebook and Twitter. I also install apps that hide photos and videos, and I shut down my laptop completely whenever not using it, instead of keeping it in sleep mode, so the operating system is not running at all.

Stefan Tanase, a security researcher at CSIS Group in Denmark, frequently travels abroad and has a simple trick if he fears a hotel or apartment might be bugged: "If you enter your room and feel something is not right, make a cup of coffee, stain the carpet, and ask for another room."

While traveling, he tries to keep his devices with him all the time. When he must leave a laptop or a hard drive in a hotel, he grabs a potato chip bag from the minibar, spreads its content on top of his devices and takes a photo. If someone walks in while he's away and goes through his stuff, they won't be able to rearrange the chips in the same manner, and at least he'll know that something happened.

Tanase also avoids connecting to a hotel's Wi-Fi or to hotspots in cafes, restaurants, or airports. He also suggests using VPN apps for both smartphones and laptops, which encrypt web traffic to make it unreadable to the owner of whatever network we connect to. VPNs are also useful for browsing in countries where certain websites are banned because of censorship rules.

In some cases, it might be useful to avoid electronics altogether when abroad. Years ago, I bought several audiocassette recorders from a flea market and find them perfect for sensitive interviews, especially when traveling. Unlike the digital files we store on our laptops or phones, audiocassettes cannot be instantly copied or deleted when passing through border control, for example.

Google Yourself

Many journalists try to protect themselves against powerful organizations with plenty of resources, but they can also be targeted by readers who have strong opinions about stories, as Meghan Rosen, a former staff writer at *Science News*, found out.

In 2015, she wrote a seemingly benign piece about ticks, soon after her baby got bitten by one. She wanted to potentially help other mothers, telling them that ticks are not as bad as they are portrayed. Rosen interviewed several experts who explained that not all ticks carry Lyme disease, and that antibiotics can often treat it. Some readers disagreed with the researchers and flooded the comments section of the story. One person in particular posted Rosen's home address online, inviting people to mail her Lyme disease ticks.

As soon as this happened, the website's moderator shut down the comments section, but to Rosen, the idea that someone knew where she lived was unsettling. She wondered where that person could have discovered her address and also worried about the information she posted online in the previous years.

This type of hack is called doxxing because early attackers usually created a .doc file where they put together all the information they could collect on a person from various sources.

Sometimes doxxing can be followed by swatting, in which police officers descend at someone's address after receiving an alarming phony call. Former *Washington Post* technology reporter Brian Krebs was among the first journalists to be swatted in 2013, after he angered a group of Russian-speaking hackers. He says at least a dozen officers with dogs and drawn weapons came to his house, after being falsely notified that a murder took place, and handcuffed him. Luckily, Krebs had filed a report with the police 6 months earlier, saying he was a journalist covering a sensitive topic and something bad might happen to him. After reminding the officers of his report, they realized their mistake and apologized.

To prevent doxxing and swatting, journalists should google themselves and their family members regularly and investigate their digital presence just like they investigate the people they cover. Security researchers also suggest deleting all the information we don't want others to see and killing accounts or subscriptions not used in a long time.

And remember that we leave behind not only our data, but also our metadata: Websites collect piles of information about our online presence and can hand it or even sell it to others who might to use it to profile us. Wolfangel and I often talk about the lack of privacy in the digital world, and despite our best efforts to use browsers that collect less information, coupled with extensions that further increase privacy, we still feel we live in a glass aquarium.

The Takeaway

Hackers are constantly finding new ways of attacking their victims, forcing security experts to scramble to keep pace to protect users. As journalists, we might find ourselves in the crosshairs. Luckily, there are ways to stay safe, with vigilance and tools like password managers, antivirus software, multifactor authentication, encryption, burners, and even a bag of chips.

Such tools not only increase our security but also can help us gain the trust of sources who have sensitive material to offer. And quite often, these are exactly the kind of people we want to talk to.

19

The Public Information Machine

James Glanz

James Glanz grew up in radio and television stations as the son of a sportscaster and DJ in the Midwest; he was writing occasional copy, hunt-and-peck style, by the age of 12. Dead broke in college, he oversold his electronics skills and talked his way into a job at a physics lab. There, he began doing his own experimental research and writing papers that landed him at Princeton, where he earned a PhD in astrophysical sciences. He then turned to journalism, and at Science *magazine covered topics like the discovery of dark energy in the universe and the doomed original design of the International Thermonuclear Experimental Reactor. At the* New York Times, *he covered the collapse of the twin towers on September 11, 2001, and spent 2 years reporting from Ground Zero. In 2003, with Eric Lipton, he published* City in the Sky: The Rise and Fall of the World Trade Center. *He has also been the* Times's *Baghdad bureau chief and has covered topics ranging from the Columbia shuttle disaster to the coronavirus pandemic.*

In science and technology reporting, having direct access to researchers is particularly useful—and deeply fascinating—because it's their job to uncover the raw, immutable truths of physical reality, from the age of the cosmos to the structure of a virus or a bridge.

But as you may already know, it can often take a good deal of effort to have a frank, direct conversation with research scientists, engineers, medical doctors, public health officials, or other experts with important expertise or unique insight into the news of the day. That's because, as with almost any other field in the modern age, these experts are invariably part of an organization—a university, a research institute, or a government agency—that seeks to manage, monitor, and control the flow of information to the public.

In the sciences, the information control apparatus has some features that are common to almost any industry, including personnel whose job it is to handle press inquiries, manage what information is shared (and what's not), and otherwise ensure that the institution is placed in the best possible light.

James Glanz, *The Public Information Machine* In: *A Tactical Guide to Science Journalism.* Edited by: Deborah Blum, Ashley Smart, and Tom Zeller Jr., Oxford University Press. © Oxford University Press 2022.
DOI: 10.1093/oso/9780197551509.003.0020

These handlers, often called public information officers, or PIOs, are a fact of life for journalists. Learning to work with them (or in some cases, around them) in order to get access to unvarnished details and insights will be necessary if you aspire to journalism in the public interest.

In covering the sciences, you'll also need to grapple with a concept that is especially prominent in the universe of published research: the embargo. This is essentially a set of rules, typically imposed by a research journal ahead of publication of a new study, governing precisely when a journalist can report on the findings. The journal and the scientists will give you and your colleagues an advance look at select details, in exchange for your agreement not to publish on the topic until a date and time they establish.

Embargoes exist in other areas of reporting, of course, but the authority they have assumed in scientific research is hard to match. The core of this authority has been maintained and enforced by journals founded in the 19th century, and if a writer breaks an embargo, that can result in termination from mailing lists, angry letters to editors, and sometimes public naming and shaming.

Faced with this obstacle course, reporters may feel that they have few choices. They can take it easy, give up their efforts to find a scoop or produce truly independent reporting, and hope their graceful writing skills let their articles rise to the top when everyone is writing the same story at the same time. But if you find that dispiriting, there are other options: You can develop a radar for finding and cultivating relationships with press handlers, turning them into allies in the service of journalism. You can find ways to get around embargoes that don't leave you blackballed by the information control machine. And you can develop strategies for getting scientists to speak directly and candidly to you, and—crucially—to feel they made the right decision when the story appears.

Choosing to reach beyond the information control machine requires having a tough skin and a willingness to push against forces that are designed to hold you back. But it comes with far greater professional satisfaction—and the knowledge that you are working on behalf of readers and not at the behest of the people and institutions you cover.

What to Expect

Scientists do not want to be seen as enforcers. They want to be considered teachers, guides, benevolent uncoverers of the laws of nature. In so many ways, they are, and the task of a journalist is to find ways to tap into that underlying character. But don't kid yourself: Scientists will go so far as to sue you if they don't like what you write (don't take this lightly), they will ignore your deadlines if they are not sure providing comment is good for them, and they are well aware that the public relations (PR) apparatus is there because they back it.

The first time I got this lesson, when I was writing for the news hole at *Science* in the mid-1990s, it came as a one—two punch. I had reported on the flaws that a pair of young physicists discovered would keep a $10 billion nuclear fusion device from producing any useful energy. The PR shops at several major research institutions circulated a press release, "Fusion Scientists Question Validity of Science Magazine Article." An irate physicist on the project then called and bellowed so uncontrollably that I had to hold the receiver 2 feet from my ear just to make out what he was saying.

Hearing directly from the angry scientist was the eternal value of this lesson. It's never just some faceless press release that is trying to undermine your reporting or damage your career: Scientists, in the most charitable view, allow it to happen, or at least do nothing to stop it. How do you grapple with that kind of attack? Force the scientists to own up to it. I contacted the first scientist listed at the bottom of the press release—by implication, someone "questioning the validity"—and he said that he had never agreed to be included in the release. Within the next day or so, the same PR shops did not quite issue an apology (I am still waiting for that) but backed off their absurd attack on an article that was correct.

Cracking the Embargo

Before becoming a reporter, I had a short but enjoyable career as a physicist. I had never heard of an embargo before becoming a reporter, and the custom struck me as something out of a Monte Python comedy sketch. In the bizarro world of embargoes, all reporters agree to study a layperson's version of a research paper, perhaps confidentially reach out to scientists outside the project for comment, and then, in unison, publish their stories on the research finding at the same moment—say, "8 p.m., Wednesday, April 21."

It's true that doing journalism this way is the rough equivalent of pretending to ride horses as someone trots behind knocking coconuts together to simulate the sound of hoofs striking the ground. What makes it worse is that its dominance among established journals has given license to every university research department to pass out flurries of embargoed releases about every gee-whiz finding their scientists have written up.

If you have ever tried to get basic physics research on Page 1—my mission at the *New York Times* before 9/11 scrambled our lives and careers—then you know that roughly three elements are essential to have any shot at all. The story has to reveal something about the wondrous nature of physical reality and its magical ability to reveal things that seem, by any ordinary estimation, impossible; the evidence must be as strong it can be; and, with rare exceptions, you need to have an exclusive.

So I learned to find ways around the embargo. I'd go to conferences where scientists presented their work and reported what they said, even though the full paper might be under embargo. If there was an embargoed paper I wanted to write up, I'd look for a similar paper set to be published at a journal without an embargo policy and try to persuade the embargoed journal to drop its restrictions in order to be a part of my article.

That was the case in January 2001, when two groups of physicists had slowed light from its usual 186,000 miles per second to exactly zero—light at a standstill!—and let it fly free again. Because of, for lack of a better term, the incredibleness of the finding, I would need both papers for my editors to deem it believable. And because the embargo was looming, I needed to file that day to be sure of the scoop and have a shot at Page 1.

One of the papers was embargoed; the other (though no one else had noticed it yet) was not. But because the embargoed journal initially refused to play ball, I walked over to my editor and explained the situation. She got on the phone with the embargoing journal. Voices were raised. The journal finally agreed not to cause a ruckus if we cited the paper—but, absurdly (can you hear coconuts?), would not allow me to quote the researchers. So be it. We had our scoop.

The power of embargoes has in some ways intensified since then. My theory is that there's constant turnover among science journalists because new writers play along for a while until they figure out the rules and leave the beat in frustration.

A few cautionary words: Don't let yourself be trapped when journals rely on "peer review"—the practice of recruiting a few outside experts to assess findings anonymously before deciding whether or not to publish a paper. What matters is whether the results seem solid according to the experts you can muster and by your own lights.

In the story about stopping light, I cited the kind of source every reporter should have in a deep contact list, built up over years: scientists who can help vet the quality of research they were not directly involved in. People who are really, really, smart and offer great quotes as well are excellent additions to that list.

A coda: Preprint servers that allow researchers to post article drafts before publication do not solve all your problems with access or with embargoes. Journals know the preprint servers exist, too, and if they have a strict policy on embargoes, they will not allow the research to be posted there. On the other hand, some journals have made their peace with researchers who post on servers to provide the scientific community with quick access to their work, so some embargo policies are essentially formalities.

Seek Out the Best Gatekeepers

While there is a shortage of capable, helpful press handlers, they do exist. Some are, in fact, spectacularly good. You need to find these outliers and cultivate relationships with them. Many are former reporters; others began their careers in bare-knuckles areas like politics where reporters don't play as nice; still others are just independent-minded professionals who know that getting good placement for a science story is more important than bureaucratic roadblocks.

It would be impossible to count how often I have leaned on these pros. One was instrumental in the scoop a colleague and I wrote about a confidential report by the U.S. Nuclear Regulatory Commission warning that the Fukushima Daiichi nuclear plant, which had been struck by a tsunami, faced new threats "as a result of the very measures being taken to keep the plant stable." Several others helped with scoops on the phylogenetics and infectious disease modeling of coronavirus in the annus horribilis of 2020. And another worked behind the scenes on the story about stopping light to ensure full access to the unembargoed paper.

The most straightforward way to remove PR officials from the equation is to go directly to the scientists. Like generals in the military, scientists usually have enough stars (metaphorical ones) on their shoulders to bypass their handlers if they want to. But this approach, too, comes with its prerequisites and perils.

If you have ever covered a war, you know that a little understanding of military culture goes a long way toward getting you to the points of the compass you want to report on. The same holds for scientists. They may be woolly-headed intellectuals, they may be buttoned-down academics, or they may be nerds who wear the same stained shirt every day of the week. (I knew a senior physicist like that.) But they share the same culture.

Some of that culture is obvious; some is not. Scientists want you to show respect by doing your homework before the interview. Scientists, by and large, do not want to be the focus of a puff piece that would make them look foolish in front of skeptical colleagues. Most of all, scientists do not want to be accused of publishing their work in what they like to call "the popular press" before it appears in an academic venue.

So it pays to be not only persistent but also creative. While working at *Science*, when I received a tip that a team of supernova chasers was going to turn the cosmos as we knew it upside down, showing that two thirds of the universe consisted not of matter but of a spongy, ethereal substance called dark energy, I worked out an arrangement with the scientists: I wouldn't post my story until one of their lead researchers began a seminar talk on the findings in Marina del Rey, California. In return, they gave me full access to the findings and the scientists beforehand.

Chaos ensued, as it usually does after big stories. In the end, though, it all seemed to work out. Not for lack of trying, no one was able to disprove the existence of dark energy, and a total of three researchers on two independent teams won the 2011 Nobel Prize in Physics for the discovery.

The Takeaway

Something I've always kept in mind is that scientists, as inspiring and essential as they are, have damaged two important fields by creating and backing what is now an ossified information control machine. The first is science reporting, by driving a constant brain drain of talented people who don't see the invented rules as legitimate journalism. The second is science itself: Many readers know if a finding has been given a hardball examination or is a piece of puffery, and they also know that a lot of puffery is later shown to be wrong.

The current lack of confidence in science among the public has many sources. One of them, whether scientists want to believe it or not, is the often shaky reporting that comes out of the information control machine. So never apologize for throwing a monkey wrench into the works.

PART IV
COVERING SCIENCE BEATS

20

Medicine

Sabriya Rice

Sabriya Rice is the Knight Chair in Health and Medical Journalism at the University of Georgia (UGA), where she helps train the next generation of students interested in writing about health, medicine, and science. Prior to joining UGA's College of Journalism and Mass Communication, she reported on healthcare and medicine for a variety of media outlets. She was the business of healthcare reporter for the Dallas Morning News, *covered quality and safety issues in U.S. hospitals and health systems for* Modern Healthcare, *and produced television and digital medical news stories for CNN.*

One of my first bylines on the medical beat was in 2006 while I was fairly new to CNN's medical unit. Late one Wednesday afternoon, we learned of a new study being released in the journal *Nature* in which researchers would detail an enhanced map of all the genes that make up the human body.

It was a big deal. But I was petrified.

I only had a few hours to read and comprehend the study, interview the researchers, and find an expert to offer an independent perspective. Plus, I'd need to write a TV script for the medical correspondent, Judy Fortin, who would go live on air to explain the findings to the public.

The study was full of phrases like "delineate linkage disequilibrium" and "single-nucleotide polymorphism" that, to me, read like a foreign language. I was not only worried about sounding stupid on the phone with the study's authors, but also concerned about reporting the findings incorrectly in the final story.

It turns out that the anxiety I felt is common for newbies on the beat. "Everything made me nervous," recalls Felice Freyer, a long-time health reporter for the *Boston Globe*, when asked to reflect on how she got started. "I was terrified."

The journalists who cover healthcare, medicine, and science help the public make sense of complex information from major industries. They cover medical discoveries, help explain sweeping changes in healthcare laws, instruct people

Sabriya Rice, *Medicine* In: *A Tactical Guide to Science Journalism*. Edited by: Deborah Blum, Ashley Smart, and Tom Zeller Jr., Oxford University Press. © Oxford University Press 2022. DOI: 10.1093/oso/9780197551509.003.0021 on how to decipher big medical bills, and delve into a wide range of health-related topics, from fitness to hospitals. For medical reporters, these stories can challenge you intellectually and often provide a natural human-interest angle, while deepening your understanding of business, science, and government.

Coverage of the Covid-19 pandemic has demonstrated that nearly everything in our lives has a potential healthcare angle. "Any story could be a healthcare story," says Atlanta-based freelancer Max Blau, whose stories have crossed into beats such as education, business, and politics.

But learning how to cover them can be daunting.

What I have learned over time is that the skills that generally make good journalists—such as developing strong sources, clear writing, and being curious and persistent—can help any reporter navigate this beat. Those skills, plus knowledge of certain best practices, can make the reporting experience less disconcerting for new journalists suddenly assigned stories that involve patients, hospitals, medical studies, and, likely, a host of medical jargon.

Don't Be Afraid of Data

One thing people notice immediately when they delve into medical reporting is the abundance of data. There's a huge range—from the prevalence of disease in a community, to the newest drugs and therapies, to how much money local health systems spend.

When you begin health reporting, it's a good idea to identify what type of health information is being tracked in your state or region and figure out how often the data are released and in what format.

That was essential when I began writing for the *Dallas Morning News* in 2016. Subscribers began reaching out to me expressing their frustrations with extremely large medical bills. In emails and on social media, they told me about the high price they paid for visiting freestanding emergency rooms (ERs) in North Texas. Those privately operated emergency centers were not physically attached or affiliated with full-service hospitals and were often not a part of any health insurance network. Consumers often confused them with lower cost urgent care centers—that is, until they got the bill.

I quickly got up to speed on who owned and operated the facilities. I learned what data the ERs had to report to state and federal officials. And I identified what fell within the public's *right to know*, meaning the information that should be publicly accessible. I submitted public records requests and created my own spreadsheets. I could then track how many times patients contested their medical bills and see how often freestanding ER staff had to call 911 to transfer the patient somewhere else.

That prompted a series of stories that revealed a lot about the logistical and financial struggles of Adeptus Health, one of the largest operators of these types of facilities before the company filed for bankruptcy in 2017.

My experience in North Texas resonated with MaryJo Webster, the data editor for the *Star Tribune* in Minneapolis. She works with reporters across many different beats in her newsroom and also participates in programs that train reporters in working with data.

She notes that those who write about health and medicine in her current newsroom are often among the savviest when it comes to data collection. "They learned data because they realized they needed to as health care reporters," she said. Those reporters still ask for help with major projects, ones with data sets containing millions of records or anything that needs advanced statistical analysis. But the key thing, Webster says, is that *you* can do it, too.

"It's all learnable," she said. Becoming familiar with data—even on a basic level—is essential to success.

Reporters who cover health, medicine, and related sciences should seek opportunities to collect data and work in spreadsheets. And if you can, it's worth taking a class or two in statistics, adds CNN senior medical correspondent Elizabeth Cohen. Understanding terms like *statistical significance*, knowing what a *p* value is, and being able to determine the difference between *correlation* and *causation* are important skills to have as you advance on the beat.

"You don't want to teach yourself this stuff," Cohen says. "It may not become your life's work, but take that class. It makes a big difference."

A simple Google search will take you to the pages of several journalism groups with resources on data and use statistics, such as the Association for Health Care Journalists, the National Association of Science Writers, and Investigative Reporters and Editors. And books like *Covering Medical Research* by Gary Schwitzer and *Numbers in the Newsroom: Using Math and Statistics in News* by Sarah Cohen can help you if you're working on a current story for which data are a factor.

Connect With Readers' Lives

Focusing on the human element is key, no matter if the story is about data, the latest scientific research, or policy. "There is no story about Medicaid that doesn't ultimately affect a real person and their ability to live a healthy, fulfilling life," says Sam Whitehead, a reporter for WABE, Atlanta's NPR station.

It's easy to get stuck in the weeds and lost in jargon at first. But the best reporters on this beat remember that behind every data point is a real person. They find ways to use the human element to tell the story. "They remind listeners that, yeah, you might think Medicaid policy is boring," Whitehead says. "But it ultimately affects lots of people, some of whom you may know. There is nothing more personal than someone's health."

Understanding data and statistics is a good starting point for identifying patterns and trends. But it's critical to move beyond the numbers. "You can have all the perfect data points in the world, but if you don't have a good story to hook those points onto, it's going to get lost in the sea of noise," says Blau. "It sounds incredibly cliché, but it's the ultimate truth of this particular field."

Of course, it's not always easy to find people to share personal stories. For starters, it's important to become familiar with the privacy rule added to the 1996 Health Insurance Portability and Accountability Act, also known as HIPAA. The rule, which took effect in 2003, established safeguards to protect the privacy of personal health information.

The bottom line is that hospitals and health systems—and anyone who works with patient data—have few incentives to help journalists access information.

We'll get back to HIPAA in a minute. But, another challenge to bringing real people into your stories is that it's not always easy to get people to open up about their private healthcare challenges. Reporters must toe the line between being both persistent and empathetic.

Freyer of the *Boston Globe* says she used to constantly feel the need to keep journalistic distance and to avoid engaging on a personal level. But her perspective has evolved. "You have to be a person first," she said. "If you do that, people will talk to you more."

I have certainly found that to be true. For feature stories, I like to conduct two interviews. First, a phone call, where I can get all the major questions out of the way. Then I like to spend one-on-one time with the person, to see them in their element. People who share their stories often feel more comfortable when they get to know you. Nobody likes to feel like you're just waiting for them to give a good sound bite.

Know the Law

At the start of the Covid-19 pandemic, questions about privacy were among the most frequent for attorneys at the Reporters Committee for Freedom of the Press, a Washington-based nonprofit organization whose lawyers help journalists with legal issues.

As reporters sought data on the spread of the virus, many faced resistance from local health agencies and elected officials, who cited HIPAA as a reason for not sharing information. It happened so often that the Reporters Committee created a guide to help journalists navigate the complicated world of patient privacy.

When HIPAA was enacted in 1996, the key goal was to ensure that people who had health insurance through their job could keep it if they became unemployed. The privacy rule, added later, set national standards to protect the health information of individuals. This can include the patient's name, test results, images such as computed tomography (CT) scans and x-rays, and more.

As a result, the key thing to keep in mind is that people and groups involved in the healthcare system—like doctors, pharmacists, and health insurance companies—can face lawsuits and steep financial penalties for sharing health information that can be traced back to an individual patient.

Because of that, when a reporter makes a request for data or medical records, often the default answer is: "I'm sorry, I can't tell you anything because of HIPAA," says Adam Marshall, senior staff attorney at the Reporters Committee.

It's important to know when to push back. Whenever HIPAA is cited as a reason for withholding information, check to see if the source is indeed a "covered entity," a term that refers to those who must comply with the HIPAA privacy rule. Covered entities generally include medical offices, hospitals, health insurance companies, and the many others who manage patient data.

Sources like lawmakers, police officers, and consumer health app developers may have access to some health information, but technically they are not covered entities, so they can't cite the HIPAA law in denying your request. They may still choose not to give you what you want if you push back, but they'll have to identify a different reason.

Even if the source is a covered entity, that doesn't mean the information you're requesting is completely off limits. The law applies to any data that explicitly identify a person, like someone's name, date of birth, and details from their private medical record. It also applies if there's reason to believe the data can be traced back to an individual, which can happen if you're requesting data from a small sample size.

But most large, aggregated data sets are de-identified, meaning the information can't be traced back to a given person—including data on how many people were admitted to the hospital in a given year or how much people spent out of pocket on healthcare costs. This type of information is often used for research or shared with elected officials and lawmakers.

And the privacy rule doesn't apply if an individual patient gives you permission to speak with their providers. I once interviewed a man who was stuck with a $128,000 medical bill that his health insurance provider refused to pay. Both the hospital and insurer were covered entities, and as expected both cited HIPAA when they declined to comment for the story.

But I pushed back. The patient even wrote a letter saying he wanted them to speak to me about his case. Both companies ultimately agreed to speak on the record. Oh, and they also managed to settle their dispute and pay the man's bill, a few days before my story was published.

Trust, But Verify

One thing CNN's Cohen encourages interns and new reporters in the network's medical unit to do is be vigilant about double-checking information, no matter how reputable the source who provided it.

She recalls a story from early 2020, in which she wrote about the asymptomatic spread of SARS-CoV-2, the virus that causes Covid-19. The virus was just beginning to circulate in the United States, and some senior health officials she spoke to disputed the idea that people without symptoms were a major source of the disease.

"Boy, did people get mad at me—I caught so much flack," Cohen recalls. "What do you mean it spreads asymptomatically? That can't be right," she recalls some experts saying when her story was broadcast. But, Cohen had cast a wide net and done her research.

Some of the people she had spoken to were experts in viral respiratory diseases like severe acute respiratory syndrome (SARS), which was used as an initial comparison to the condition we now know as Covid-19. But others had a more nuanced perspective. Research from the Centers for Disease Control and Prevention later found that at least half of new infections could be traced back to people who were exposed to others not showing symptoms.

Speaking with doctors, scientists, epidemiologists, policy wonks, and others with deep knowledge about a topic can be intimidating for young journalists—but it's crucial. The novel virus was unpredictable, and breaking news is a situation that is always evolving.

"There's always the possibility that someone could be wrong, no matter what their position," Cohen says. "Speak to many different scientists, because they often have different points of view."

It's important to question every assumption, especially the ones you're most comfortable with.

The Takeaway

If you're nervous about health and medical reporting, don't be. Everyone has a health story. The pandemic proved that there's a place on this beat for every journalist, whether you're interested in education, sports, policy, or science.

Just remember, there's no need to overthink the content. Ask yourself: *What are the one or two things I want my audience to learn?* and stay focused on that. Look at data yourself—even if that means asking someone for their medical record—and make sure each number used in the story makes sense for the audience.

And always be respectful when dealing with patients. "People can sense when you're looking to just get a quote and go on to the next story," says Blau. "When I'm asking someone to put themselves on the line, I want them to be as bought into the story as I am."

I also keep in mind my own friends and family, whose everyday experiences in the healthcare system helped me identify potential story angles and stay focused on real people.

21

Infectious Diseases

Helen Branswell

Helen Branswell is the senior writer on infectious diseases and global health at STAT, a role she has held since the site's launch in the autumn of 2015. Prior to joining STAT, she was the medical reporter for the Canadian Press *for 15 years, based in Toronto. In 2004, she spent 3 months embedded at the Centers for Disease Control and Prevention (CDC) as a CDC Knight Fellow. In 2010–2011, she was a Nieman Global Health Fellow at Harvard, focusing on polio eradication. She won the George Polk Award for Public Service in 2020 for her coverage of the Covid-19 pandemic. She had no idea she would one day be a science writer when she earned her bachelor's degree in English literature.*

The news became more frightening with each passing day. A new disease, one that left people profoundly ill and struggling for breath, had broken out in China, with cases linked to live animal markets. Details were sparse and grudgingly released, though it was rapidly clear that what started there had made its way elsewhere. Global travel was quickly restricted, cratering economies. Hospitals were closed to all visitors, forcing those at the end of life to die alone.

This may sound familiar, but it's not what you think. This is the story of the start of the 2003 severe acute respiratory syndrome (SARS) outbreak, the event that hooked me on covering infectious diseases.

I was a health reporter for the *Canadian Press* based in Toronto, newish to the beat and completely ignorant of the ways of infectious diseases. But when Toronto became the only place outside of Asia to experience a full-blown outbreak of SARS I was given a full-on crash course on reporting in an epidemic. It was months of 7-day workweeks, fueled by adrenalin and caffeine.

I wouldn't wish another SARS or Covid-19 (caused by a cousin virus, SARS-CoV-2) on the world. But these and many lesser disease outbreaks are going to happen. As ghoulish as it sounds, covering these events is one of the best jobs science writing has to offer.

Helen Branswell, *Infectious Diseases* In: *A Tactical Guide to Science Journalism.* Edited by: Deborah Blum, Ashley Smart, and Tom Zeller Jr., Oxford University Press. © Oxford University Press 2022. DOI: 10.1093/oso/9780197551509.003.0022

People are mesmerized by infectious diseases. Advances in cardiac care or cancer treatment will doubtlessly save more lives, but unless you, a close friend, or a family member are struggling with one or the other, stories about those developments may not draw your eye. But almost everybody wants to read about Ebola, even though most of us will never be within catching distance of it. In a bad flu season, a story explaining why influenza vaccines consistently underwhelm will always find an eager audience.

An Inexhaustible Beat

I like to learn a lot about a topic I'm going to write about. Going deep helps me find stories other journalists aren't reporting, and the infectious diseases beat is ideal for this. There's enough variety to keep things interesting, but not so much that you're always just skimming the surface. Plus, every few years, Mother Nature tosses us a curveball: a flu pandemic, caused by H1N1, in 2009; a SARS cousin, a camel virus called MERS (Middle East respiratory syndrome) that occasionally infects people, in 2012; a massive Ebola outbreak in 2014; and Covid-19 in 2020.

Perhaps more than any other kind of science reporting, covering infectious diseases is often about reporting on breaking news. This news includes Ebola making its way from West Africa to hospitals in Dallas and New York City; emergency rooms buckling under the crush of patients during a norovirus outbreak; and public health authorities issuing vaccination orders to try to quell a mushrooming measles outbreak.

The cadence of this kind of reporting can be intense. Your deadlines will generally be tight, in keeping with the pace at which many of these outbreaks unfurl. Beautiful prose will always earn you plaudits, but sometimes speed is the priority. Being able to report and write quickly, clearly, and accurately is a skill you'll never regret acquiring.

Learning how to read an outbreak—is it on the cusp of going next level or coming under control?—is also critical. Some are explosive, fueled by the exponential growth that made Covid-19 a pandemic. Others grow at a different pace.

These events affect economies and sometimes topple governments. Donald Trump's mishandling of the Covid-19 pandemic contributed to his failure to win a second term in 2020. Both SARS-1 and SARS-2 had enormous impacts on the global travel industry. Outbreaks also have roots in environmental issues; incursions into the planet's forested preserves led to the spillover of bat and rodent viruses into people. Nipah virus outbreaks in South and Southeast Asia are such examples, and the recent increase in Ebola outbreaks is another.

How do you learn enough to cover them? Unless you have an advanced degree in microbiology, you do it the way I did. You do a lot of reading and ask a lot of questions.

Getting Up to Speed

When SARS hit in 2003, I knew nothing about it. No one else did either. It was a new disease, the cause of which was unknown. The science of SARS evolved in real time. Within weeks (fast back then), a new coronavirus was identified. As scientists raced to probe its mysteries, I struggled to learn the basics of infectious diseases: how they spread; what, when, and who they infect; and how quickly they transmit and how rapidly some outbreaks expand in comparison to others.

Some of this you'll pick up on the fly, finding out at the same time about the general concept of incubation periods and the specific incubation period of the disease you are facing. Are people infectious in the prodrome—the period between when a person contracts a pathogen and when disease symptoms start to manifest—as is the case with Covid-19? Or do infected people only transmit after they get ill, as was the case with SARS-1? These are crucial questions that will give you an idea of how containable or not a disease is.

With every outbreak you cover, you'll learn more about the characteristics of diseases, features that contribute to how well and how rapidly they spread and who is at risk. Figuring out what you need to know to understand any pathogen will flatten your learning curve for the next one you encounter.

Knowing where to look for information helps. The website of the Centers for Disease Control and Prevention (CDC) has an A-to-Z index where you can research everything from arenaviruses, transmitted from rodents and pretty dangerous, to Zika, inconsequential except in rare circumstances. PubMed, a search engine for published scientific papers maintained by the National Library of Medicine, will help you find review papers on various pathogens, another good way to get up to speed. PubMed will also help you figure out who is expert in a field and how to find them. If I can't locate an email address for someone, I search for them in PubMed. If they are the first or last author on a paper, chances are their contact details will be listed.

Other useful resources are as follows: The European Centre for Disease Prevention and Control (ECDC), publishes informative, readable risk assessments of disease outbreaks. The World Health Organization (WHO) issues updates known as DONs, short for Disease Outbreak News, on events of note around the world. Countries are obliged to report serious disease outbreaks to the WHO, so DONs will tell you what the agency is hearing through official channels.

The U.S. CDC publishes two online journals that are a good source of outbreak information: *Morbidity and Mortality Weekly Review*, which is where much of the earliest epidemiological information about a new outbreak will be published, and *Emerging Infectious Diseases*, which is more like a standard journal but focused on diseases that cause outbreaks. And do sign up for ProMED, an infectious diseases listserv run by the International Society for Infectious Diseases that played a key role in alerting the world to SARS-1, SARS-2, and myriad other outbreaks over the years.

My profile on STAT's website used to warn about my propensity for asking lots of questions. The deeper you go, the more likely you'll turn up another story that can follow the one you're working on. But read the room; keeping someone on the phone for longer than they budgeted for the call may make them less inclined to agree to an interview the next time you ask. And during an outbreak, people you want to talk to will have many demands on their time. In those cases, ask the questions you really need answered, thank them, and let them go.

Finding the Right Experts

As in any type of science journalism, you will rely heavily on experts. But the generic term "infectious diseases expert" covers a lot of turf.

Microbiologists and virologists study the genetic structures of the pathogens about which you are writing and how they function. They may work on multiple pathogens, but typically will have a focus, whether it's flu or HIV. Others will be immunologists, who study the workings of the immune system, or vaccinologists, who develop and test experimental vaccines.

Some infectious diseases experts practice public health; many will be epidemiologists or "disease detectives." Others will be mathematical modelers, who use complex formulas to figure out things like how many people need to be vaccinated to stop transmission of a pathogen—in other words, to reach herd immunity. Some are evolutionary biologists who track the spread of pathogens by studying their genetic sequences; they can estimate from those sequences how long a new virus that had an animal source has been circulating in people. Some are infection control practitioners who work to keep hospitals free from outbreaks, including those caused by antimicrobial-resistant pathogens.

You want and need to get to know as many of these folks as possible, from the full range of professions in the field. You also want to make sure you are talking to the right kind of expert for the story on which you are working.

To that end, here are a couple of things you should know.

No one knows everything about infectious diseases. A flu expert isn't the right person to talk to when local children have a brush with a kitten that later tests positive for rabies. Look for people who have published on the topic you are writing about. If someone is being pitched to you as an expert, check out what they have published. My trust level always rises when someone tells me: "That's not my area." I'm more likely to circle back to them when I'm working on a story that *is* in their wheelhouse.

A major outbreak will always draw would-be experts out of the woodwork; with every epidemic, a cadre of people become cable TV stars. Some of them are actually deeply knowledgeable about the issue at hand. Some are people with reasonable credentials who know how to read scientific papers and craft sound bites. If you are just looking for a quick quote—the same quote everyone else will get—the last group can give you what you are looking for. If you're looking for true insight and intel about what the CDC or the WHO or the leadership of your country is considering, these people are unlikely to provide it. They synthesize other people's information.

Here's a tip on how to detect the true experts from the would-be experts: You will rarely get a press release offering you interview time with someone who is a true expert when an outbreak is underway—or at any other time. (There are exceptions: A university might offer up its best people, or the Infectious Diseases Society of America might stage a briefing featuring legitimately valuable people.) People who are true leaders in a field typically do not hire public relations firms to book them interviews. They don't need to.

Silver Bullets Are Rare

In the mid-aughts, when the world feared that a potentially catastrophic bird flu pandemic was in the offing, the limited global capacity to produce influenza vaccine was a huge concern. An academic who was a prolific writer of journal commentaries came up with what appeared to be an ingenious solution: Add an adjuvant, a boosting compound, to the vaccine to stretch supplies. Alum—potassium aluminum sulfate—is a cheap and safe adjuvant that had been used with vaccines in the past. This would stretch supplies like the biblical loaves-and-fishes miracle, he argued. There was just one problem. It didn't work.

During the devastating 2014 Ebola outbreak in West Africa, some scientists contended that silver nanoparticles or statins, commonly used cholesterol drugs, would reduce the death rate among infected people. Early in the Covid-19 pandemic, an antimalarial, hydroxychloroquine, was heavily promoted to both treat and prevent infection—including by the U.S. president. Both would have been terrific solutions to pressing needs, if they had been effective. They were not.

It makes perfect sense to scour the medicine cabinet when a new disease arises. If an existing drug works, repurposing it saves time and money. Drugs that have already come through Food and Drug Administration (FDA) licensure are being produced, meaning they are instantly available.

There are occasional successes. Dexamethasone, an inexpensive steroid, was shown to lower the risk of death of severely ill Covid-19 patients. But more times than not efforts to turn existing drugs against a new enemy will fail. Despite that, a remarkable amount of hype will sometimes be generated about the hoped-for silver bullet.

Apply a skeptical eye to these efforts. Exuberant coverage of unproven "therapies" could prompt people to take drugs that won't help them and may hurt them. And there can be other negative consequences. When hydroxychloroquine was hyped for Covid-19, people with lupus who use it to control their symptoms had difficulty accessing a drug they badly needed.

View claims of miracle cures in an emergency with the same jaundiced eye as you would in normal times. Review the data. Talk to experts. Don't amplify the hype.

Expect the Unexpected

The longer you cover infectious diseases, the more you'll understand how these events unfold. But even years of experience can mislead you. It even happens to the experts.

In early 2014, Ebola broke out in Guinea, a West African country with no experience handing the deadly disease. The WHO and the CDC deployed experts to help; by early May, transmission appeared to be petering off. Even long-time Ebola field experts thought the outbreak was coming under control.

Instead, it had gone underground, with infected people fleeing to evade detection. Within weeks, Ebola erupted in Guinea's capital, Conakry. It was the first time Ebola took root in an urban setting, and the outcome was a disastrous epidemic that engulfed three countries, ended 11,000 lives, and took 2 years to contain.

Likewise, the world's expectation that coronavirus outbreaks could be controlled—based on the SARS epidemic and occasional flare ups of MERS—led some experts to initially downplay Covid-19. Several told me in January 2020 that they believed the new virus would not trigger a pandemic. But the virus, SARS-CoV-2, turned out to be a vastly more difficult to control spillover agent than SARS-1 because of differences in the transmission dynamics.

Some experts also predicted that the SARS-2 virus would not mutate in significant ways because that's not what coronaviruses do—or so they believed. But with free range to infect hundreds of millions of people around the globe, the virus began developing variants that were more infectious or escaped the human immune response.

The lesson is clear: There are always new things to learn about old bugs, and newly discovered cousins of known agents may not behave in the ways you would expect. It's important to question your assumptions when you think you have detected a pattern based on previous evidence.

The Takeaway

To do this type of reporting right, you need to learn the basics, keep an open mind, and speak to the right people. Dive in as deeply as you can. And be skeptical when someone claims to have found a silver bullet.

22

Public Health

Julia Belluz

Julia Belluz is Vox's senior health correspondent. She's spent more than a decade reporting on medicine, science, and global public health across platforms and media. Before joining Vox, she was a Knight Science Journalism fellow at the Massachusetts Institute of Technology. Her writing has appeared in a range of international publications.

Reporting on public health can be intimidating. Your sources tend to be doctors or researchers who don't speak in plain English, patients who have been harmed by devastating illnesses and shortsighted health policies, or politicians who spin science. The subject area is also vast. In the space of a few days, you might swing between covering the quick-breaking news of an Ebola outbreak, a report on the dangers of drinking, or new e-cigarette regulations.

The learning curve is steep, and the stakes couldn't be higher: You're often covering fundamental questions of life and death and how societies care for people and share resources. More than many other areas of science journalism, public health is also political. It's as important to understand the politics and interest groups that might be shaping policy and regulations as it is to grasp the science behind an issue.

But with some time on the beat, it'll be clear that it's immensely rewarding. You learn new things all the time, and they're particularly useful things since health matters to everyone. You can help your audience better understand themselves and the world around them. Your work might even influence political decision-making.

To help you get started, and maybe save you some pain, here are some of the most important things I've learned in my years of reporting.

When It Works, Public Health Is Invisible

There are a few basic facts that you need to absorb when starting on this beat. The first and most important: It's focused on preventing disease and keeping

Julia Belluz, *Public Health* In: *A Tactical Guide to Science Journalism.* Edited by: Deborah Blum, Ashley Smart, and Tom Zeller Jr., Oxford University Press. © Oxford University Press 2022. DOI: 10.1093/oso/9780197551509.003.0023 people healthy, which means it's mostly invisible when it's working—a non-story. When a community's HIV prevention strategy stops the virus from spreading, for example, you won't see cases rising. When a municipal water supply system properly filters drinking water, children won't be sickened by lead. When a state has a robust vaccine program with few loopholes, vaccine-preventable outbreaks will be a rarity. But news emerges, and you get busy, when these systems fail: HIV cases suddenly spike, kids are showing signs of lead poisoning, or there's a measles outbreak.

Public health also operates on long-term horizons—planning for outbreaks of yet-to-be-identified pathogens, curbing smoking to prevent tobacco-related disease, or finding ways to encourage people to exercise and eat healthier foods. The rewards of these efforts typically pay out years or even decades later. They're also difficult to quantify, especially when public health does a good job.

"Like climate change, illnesses down the road are not easy to see until they actually cause the damage," says *New Yorker* staff writer Michael Specter, who focuses on science, technology, and global public health. "And it is human nature to say, what's the big deal if you don't see it?"

For these reasons, public health is often referred to as the poor cousin of medicine—it's chronically underfunded in relative terms. Also unlike medicine, which focuses on treating disease in individuals, public health focuses on the collective, as small as a village or as big as a country or even the world.

There Are Always Costs and Benefits

Every public health intervention you'll report on comes with benefits, as well as risks, costs, and trade-offs. Politicians and policymakers are supposed to weigh them and use the best-available science to guide their decision-making—but they don't always follow the science, of course. In democratic countries, they need to consider the wishes of the electorate. Short-termism in politics and the lobbying of interest groups can also trump science—like we saw with the debates about mask wearing during the Covid-19 pandemic or America's relatively lax regulation of marketing for tobacco products. When decision-makers stray from expert advice or the best-available science, you'll want to see if there's a story to cover.

You'll also want to look out for those risks, benefits, and trade-offs and make sure you tell your audience about them. When the leaders of a country decide to seal their borders during a pandemic, disease cases and deaths might fall—and hospital systems won't be strained—but tourism will suffer, and trade and diplomatic issues will arise. Forced quarantine for a person at risk of carrying a virus might protect a community, but it will also mean that individual pays for a hotel room, stays away from work, and avoids family or friends for a couple of weeks.

Even vaccines—among the greatest public health tools of all time—work this way: They protect communities through herd immunity, but they can still carry risks for individuals. Like everything in biology, they also come with some uncertainty. "People have often [asked] me, 'Can you promise that my kid won't get sick or injured from a vaccine?'" Specter says. "And I always say, 'No, I can't promise that. I cannot ever promise anything in biology.'" But if a million people get vaccinated, the percentage who get sick and die from the disease will be much, much higher among those who aren't immunized; he added: "And we sort of have to do this based on society."

Conflicts of Interest Are Rampant

Health is an industry like any other. Money, power, and politics—again, not just science—are deeply embedded in decision-making. Over the years, I've become hyperattuned to conflicts of interest and special interest groups because they are simply ubiquitous. I've come across studies funded by the makers of paper towels that portray electric hand dryers in a dangerous light (really!); tracked the millions of dollars in funding that multinational chocolate-makers poured into research in order to help turn dark chocolate into a health food; and exposed how highly paid management consultants quietly shape the global health agenda for the world's poor.

Powerful interest groups not only influence science and decision-making, but also can shape the conversation. Covering the emergence of e-cigarettes, I tracked how the industry made what was known about the devices seem far more certain than what the actual science showed. My reporting on new soda taxes in some U.S. cities revealed the tens of millions of dollars American beverage makers invested to dissuade voters.

But while money is certainly an important driver of agenda setting, there are also nonfinancial conflicts of interest—like researchers and scientists whose careers and research programs are heavily invested in particular ideas. All of these conflicts can be a fertile place to look for stories. They're also something to report to your audience whenever possible. And don't hesitate to ask your sources about their conflicts and look out for disclosures in studies. They don't always mean a source or study isn't trustworthy or reliable, but you want to be aware of them and communicate them to your audience.

Avoid Sources With Easy Explanations

So which public health sources should you turn to? First, you'll want to figure out which kind of specialist can go deep on the story you're covering. "Know the differences between different kinds of experts," says Apoorva Mandavilli, a *New York Times* global health and science reporter. "I've seen some stories quoting a pediatrician about whether schools should be open" during the coronavirus pandemic. But school closures are not the domain of pediatrics, which is focused on delivering medical care to children. An epidemiologist, who studies disease spread, would be a better source.

I tend to avoid sources who go outside their lanes of expertise or offer easy explanations and solutions for complicated problems—which just about all health problems are. André Picard, the long-time public health reporter for the *Globe and Mail,* tries to avoid people with extreme views. "I used to always ask people, 'What's good about what your critics say?'" he says. "That's a really enlightening question to give you a sense if people are scientific in their thinking and not too dogmatic."

You'll want to hang on to the contacts who can help you see the big picture, explain things clearly, and put them in context. But as you build a roster of sources, take care not to turn back to the same people for opinions all the time. Call up or meet as many people as you possibly can and aim for diversity—including geographic, gender, and racial. These different perspectives will make your stories richer and more nuanced.

Context Is Crucial

Study the local culture and health system you're reporting on; it matters hugely when it comes to public health. The reason people opt out of the measles vaccine in an Amish community in rural Ohio probably isn't the same driver of vaccine hesitancy as among Orthodox Jewish parents in New York or Somali immigrants in Minnesota. Some parents are scared of a vaccine because of an adverse reaction that happened in their community; others have concerns rooted in their religion or were the victims of antivaccine propaganda.

The same pathogen can also behave differently in a new setting. While reporting on the 2014–2016 Ebola epidemic in West Africa, I saw how the outbreaks in Guinea, Sierra Leone, and Liberia—with their impoverished health systems and elaborate funerary rituals—spread further and killed many more people than the cases that cropped up in Europe and the United States. The same country-to-country variation played out with the Covid-19 death rate.

Also keep in mind that public health powers are distributed differently from place to place, which shapes how decisions are made and by whom. In the United States, for example, much of the public health decision-making authority falls to state and local governments. Presidential power has limits— even during a pandemic. So, for example, President Joe Biden could not order a national face mask mandate in response to Covid-19 even if he wanted to; it's not something the federal government has the authority to do. Understanding this context, including who wields the power, is crucial to getting the story right.

The Power of Data—Big and Small

Sometimes public health reporting involves covering new studies. You'll want to search for context there, too. Systematic reviews, which you can find on PubMed or the Cochrane Library, can be helpful. These are essentially research reviews that bring together the global scientific literature on a particular health question. They draw on many studies among many people at different research institutions around the world, and more heavily weight the results of the most reliable research, to come to more fully supported conclusions. They're also designed to overcome the biases inherent in single studies. Simply put, single studies are often a poor and misleading way to get an answer to a health question and often come to contradictory conclusions.

Systematic reviews can tell you where the bulk of the research is pointing. They have informed a lot of my reporting—on everything from treatments for back pain to the question of whether cell phones cause cancer.

But just like single studies, systematic reviews can be flawed and biased by design. And there are cases when single studies, or even single stories, are more relevant to the public health issue you're reporting on.

Nina Martin, a former ProPublica sex and gender reporter, realized this when she began covering the tragedy of mothers dying in childbirth in the United States. Researchers for years had been sounding the alarm on the rising maternal mortality rate in America, the highest among similar high-income countries. But the news reports that trickled out about the problem failed to move people or policymakers. Martin—along with several colleagues—decided to illuminate the stories of individual women, mining social media, GoFundMe campaigns, and autopsy reports. The result was the award-winning "Lost Mothers" project, a series of features, multimedia reports, and data visualizations that finally put names and faces to the well-known maternal death statistics. By unearthing the stories of these women, Martin not only humanized her subject but also saw patterns and trends even the researchers missed and sparked new policies to protect moms across the United States.

"What we as journalists can do is something public health people are averse to doing or not able to do because of rules," Martin says. "We can gather information. We don't have to go through institutional review boards. We don't have to go through special HIPAA [Health Insurance Portability and Accountability Act] authorizations," intended to protect medical privacy, "to reach out to a family member. We can gather information in ways that are ethical and rigorous but we are a lot freer in how we can see information."

This kind of information, despite often being anecdotal—the least reliable form of evidence according to the researchers who put together systematic reviews—can be as powerful and revealing as big studies, or even more so.

Build Communities

One way to gather a wealth of contacts and personal stories is through crowdsourcing. It can help expose information that powerful people in insurance companies, health departments, and hospital systems would prefer hidden. It can also create a community around the stories you eventually publish, point the way to follow-up stories, and most importantly, put you in touch with people you may have not otherwise found.

I worked with Martin on a reporting project that sought to explain how the broken system of health insurance in the United States was driving terrible maternal health outcomes. When we started the project, we ran a national callout through Vox and ProPropublica's social media networks and websites asking mothers to come forward and help our reporting with stories about insurance gaps they experienced that harmed their health. The callout garnered thousands of responses, and we each talked to dozens of mothers about their experiences.

With the maternal insurance project, however, it quickly became clear that the mothers worst affected by a lack of access to maternal insurance were also the most difficult to find. They tended to be living on the margins—not hanging out on social media waiting to share their stories with reporters. This experience was emblematic of one of the pitfalls of crowdsourcing in public health journalism, says Sarah Kliff, a *New York Times* investigative healthcare reporter and frequent crowdsourcer. "You don't get a lot of diversity," she says. "The people who are going to submit tend to be more educated, wealthier, and they have the free time to do that sort of thing." So, Kliff says, crowdsourcing works best "when it's a widespread experience—when individuals are holding onto a piece of information you can't get otherwise."

Martin and I had to dig deeper to find the women and families most vulnerable to gaps in access to care. That digging involved contacting doctors and health clinics that worked with patients in more marginalized communities and combing through autopsy reports to look for clues about mothers who may have died because of a lack of access to health insurance and medical care. If we didn't take these steps, we would have had an incomplete picture of the maternal health insurance problem we were reporting on.

The Takeaway

As you've probably figured out by now, public health reporting requires a lot of care and nuance. And once you master the basics, the beat gets even bigger, as Picard notes. It took years on the job for him to learn that what really matters in covering public health has nothing to do with health—at least on the face of it. Many illnesses are social illnesses.

"They have their roots in how people live," he says. "The answer isn't fancy new drugs. It's money, housing, being fed, having an education."

23

Social Sciences

Sujata Gupta

Sujata Gupta is the social sciences writer at Science News. *Prior to joining SN, Gupta completed the Knight Science Journalism fellowship at Massachusetts Institute of Technology in Cambridge, Massachusetts, where she attempted to weave science into fictional children's stories. (It turns out that writing for kids is really hard.) Before that, Gupta freelanced for a decade, writing primarily about food, ecology, and health. Her work has appeared in the* New Yorker, *NPR,* Nature, High Country News, Discover, Scientific American, Wired, Novanext, *and* New Scientist, *among other outlets. Gupta also taught journalism at Champlain College in Burlington, Vermont. She got her start in journalism at a small daily newspaper in central New York covering education and small-town politics. In previous lives, she worked as a park ranger, doing stints at Haleakala National Park in Maui, Mojave National Preserve in California, and Acadia National Park in Maine. She also taught English in Nagano, Japan.*

Sociology. Psychology. Anthropology. Economics. Linguistics. Geography. Communications. Political Science. Education. What do all these social science fields have in common? They are fundamentally about understanding people, with all their quirks and oddities, and how they interact.

By design, then, social science stories have a direct bearing on people's lives—which is equal parts boon and curse. Answering the quintessential question, "Who cares?" is easy on this beat. Insights about inequality, human behavior, and the economy affect us all. Yet, social science reporters must constantly watch for confirmation bias, or the tendency to gravitate toward stories and insights that align with their own preconceptions.

Getting this beat right matters—a point driven home by the Covid-19 pandemic. Unpacking the precise nature of the virus, and later vaccines, proved vital to containing its spread. Yet the virus spread anyway. That's partially because policymakers frequently failed to predict how people would behave

Sujata Gupta, *Social Sciences* In: *A Tactical Guide to Science Journalism.* Edited by: Deborah Blum, Ashley Smart, and Tom Zeller Jr., Oxford University Press. © Oxford University Press 2022. DOI: 10.1093/oso/9780197551509.003.0024 amid a global health crisis. Often, the task fell to social science journalists to sort out and explain what went wrong.

To say this is a tough beat is an understatement. In my few years as the social sciences writer at *Science News*, I've gotten used to covering what my colleague and behavioral science reporter Bruce Bower calls "warring tribes." Researchers do not merely disagree with one another; they disagree vehemently. Often, they define their key terms in wholly different ways, or not at all. Or they reach disparate conclusions based on how they've framed their question or the communities they've chosen to study. A case in point: One body of evidence could be used to argue that minimum wage hikes lead to job losses, while another could be used to prove the opposite.

As a result, covering social science can often feel daunting. But the process comes with inherent payoffs. Because to grapple with the social sciences is to grapple, rigorously, with the art of being human.

Digging for (Straightforward) Stories

Social science stories are often highly nuanced, and coverage hinging on a single study may be insufficient. Maybe the researchers only looked at the matter from a U.S. perspective, and a more global view would lend greater insight; or maybe multiple research teams used different terms to ask a similar question, which altered the results.

Nonetheless, reading single studies, particularly those appearing in big multidisciplinary journals and their offshoots—chiefly *Nature*, *Science*, and *PNAS*—is key to identifying some of the most impactful research in the field. Following journals within the many subfields of social sciences is also critical. But, given just how many social science journals exist, combing through every single study is unfeasible. Instead, many social science journalists hone in on a subfield or two before branching outward.

For instance, I have spent the better part of my few years at *Science News* digging into sociology. I quickly realized I should follow the dozen or so journals of the American Sociological Association. And as I gained experience and spoke to more and more sociologists, I gradually built a more comprehensive list by looking at the other journals featuring their research.

Focusing on a subfield can also elucidate the methodology unique to that given field. In sociology, researchers rarely have the option to run controlled experiments on their subjects for ethical reasons. Instead, they seek out real-life "experiments." For instance, they may compare outcomes in households in neighboring municipal regions where one region adopted a new policy—say, subsidized preschool for all residents—and the other did not.

Some of the best social science research often appears not in journals but in reports by researchers in nonprofit organizations, government organizations, or think tanks, such as the Pew Research Center, Brookings Institution, the U.S. Census Bureau, the Urban Land Institute, the Brennan Center for Justice, and myriad others. These reports can lead to great stories, but reporters must also vet such organizations to better understand their political leanings or involvement in advocacy research.

And many of the most impactful findings in the social sciences often appear first in preprint publications, such as SocArxiv, PsyArxiv, or the National Bureau of Economic Research.

A basic rule of thumb, says Cathleen O'Grady, a Scotland-based freelancer and *Science* contributing correspondent, is "to treat preprints with the same caution that we do peer reviewed research."

At a minimum, that means talking to several outside sources, often across disciplines or nationalities. Some researchers are quite comfortable discussing preliminary research. Economists, for instance, frequently post working papers online. But often, as with research emerging during the Covid-19 pandemic, researchers prefer to wait to discuss the findings until another lab has replicated the findings or the study has undergone peer review. Reporting on preprints can, as such, be surprisingly difficult. For that reason, my general strategy has been to cover preprints when they highlight broader trends, such as several research teams independently asking the same question.

Finally, many publications also offer periodic research roundups. I follow the monthly news roundup in *Behavior Scientist*, read blogs on the Society for Personality and Social Psychology website, and check Sage Publishing's Social Science Space, a valuable home to blogs, videos, and other expert commentary from social science researchers in a wide range of specialties.

Social Science and Current Events

Unlike other fields of science, which can be divorced, at least somewhat, from policies, trends, and current events, social science is tightly interwoven with the national and international conversation.

For instance, "defund the police" became a rallying cry of the summer of 2020's civil rights protests. The general concepts of defunding, such as reallocating police dollars toward social services, have appeared in research before, but only after those protests did scientists begin pursuing funding to study defunding the police as a core component of U.S. police reform. Social science journalists, as such, must stay abreast of current events to recognize the forces shaping the field's research agenda.

To that end, I keep a running list of story topics in folders in Instapaper, my online filing system of choice. Some current folder topics cover marriage and fertility trends, redlining, civil rights, and human behavior during the pandemic. The folders house links to news stories, studies, reports, videos, Twitter threads, and anything else that seems even tangentially relevant to the topic at hand. I also write notes to myself that serve as reminders for why a given tidbit caught my interest, as these folders can sit around indefinitely. My hope is that when disaster strikes or some issue rises to prominence, I can peruse those folders for ideas.

Even with diligent preparation, reporting news through a social science lens can be tricky and time consuming. I often start with a string of questions. Say, for instance, the president has proposed a new policy aimed at getting women back into the workforce. I might ask: What are the specifics of the policy? Have researchers studied a similar policy, perhaps on a smaller scale? Does a similar policy exist in another country? Is there some eye-catching statistic that is worth probing? At heart, the key question when framing a story is always: Can social science research say something about the context, interpretation, or framing of a news event?

The pandemic is, again, illuminating. Epidemiologists in the United States largely crafted the country's communications response. This approach suggested that getting the disease models right and then simply telling people how to go about their daily lives would prevent the disease from spreading. But our failure to grasp the "soft" science, such as how people behave during periods of uncertainty and the interplay between politics and human actions, no doubt contributed to our country's initial public health failure.

Cutting Through the Murk

This tendency to overlook or even belittle the social sciences arises in part because research in this field rarely presents an obvious takeaway. Whittling down human behavior during a pandemic to some key essentials, for instance, is rife with caveats. Do those behaviors vary by politics, gender, geography, or class? Have researchers overlooked some other, unmentioned criteria?

What's more, the field as a whole has struggled with a replication crisis—or an inability to reproduce findings from original research. This crisis plagues many scientific fields but has been especially prominent in the social sciences. "The reality of writing about social science is you have to go into it knowing that so much of it is poor quality," says Eliza Barclay, the science and health editor at Vox.

Social scientists have developed workarounds that can help journalists select more robust studies to cover or nail that elusive main point. Things to
look out for include whether or not the study has been preregistered, with the
hypothesis, study design, and method of data analysis stated in advance; if the
data are freely available; if the participants include more than a few dozen undergraduate students; and, if applicable, if the theory has been tested outside
of so-called WEIRD (short for Western, educated, industrialized, rich, and
democratic) countries.

If a study or studies do merit coverage, then the issue becomes identifying
how to tell the story. For instance, Barclay and her team often deal with the
replication crisis by, well, covering the crisis. She recalls an article and podcast Vox released when scientists began to question the validity of the now
iconic 1971 Stanford prison experiment, where participants assigned to the
role of guards quickly became cruel and abusive toward participants assigned
as prisoners. Vox released a story and podcast investigating reports showing
that the researchers coached the guards to be cruel and that some participants
were merely acting. Such stories have strong appeal, Barclay says, because
"there is a lot of tension. There's a lot of human drama."

Another tactic social science reporters frequently use is to look for the
gaps—the evidence that is not there. "The thing that I love the most is finding
out where the holes are in our knowledge," says O'Grady. "Also the holes help
to contextualize what we do know and how limited that is and how cautious
we should be about the evidence we do have."

A third possibility is to probe the terminology used by scientists. Early in
my tenure at *Science News*, I wrote a short feature evaluating the claim that
millennials, notorious urban dwellers, were just as likely to move to the
suburbs as the previous generation. But I quickly got lost in the research. Each
researcher I interviewed defined suburbs in different ways. And those varied
definitions changed whether or not thirtysomethings were moving to or away
from the suburbs. The story got so confusing that we wound up killing the
piece. Shortly after that fiasco, I read a news story about how measuring sub-
urban growth is impossible because of imprecision around the word "suburb."
That taught me a valuable lesson: Sometimes the ambiguity itself is the story.

Finally, while data-driven studies generate the most media buzz, reporters
should not overlook other approaches: Ethnographies, in-depth interviews,
and historic papers can provide vital context for any story. Those sorts of
studies can also provide anecdotes and narrative arcs that are crucial in feature writing. If in reporting a story, you find a mix of qualitative and quantitative research, you may have struck the jackpot.

Nailing a story's structure and framework enables social science journalists to grapple with nuanced topics without losing their readers, says Jesse Singal, a contributing writer at *New York* magazine and author of a forthcoming book on the social sciences and the field's replication crisis. "My job is to give readers complexity and give them the tools to work through that complexity."

Interrogate Your Biases

We all come from somewhere. And that somewhere defines how we construct reality. The purchasing patterns of a rich person born poor may still reflect their impoverished upbringing. A dark-skinned person born in, say, Nigeria, may not learn to identify as "Black" until they arrive in the United States. A person born and raised in a big city does not intuitively grasp the lived experience of a person born and raised in a rural town with 500 people.

These realities are true for all journalists. But in social science reporting, it's especially important to recognize that one's personal framework is narrow. That would be less of a problem if journalists hailed from diverse backgrounds. Instead, most represent a very small slice of the population, one that is predominantly White, affluent, and coastal. That doesn't mean that social science journalists should avoid contentious material, but it requires continuously interrogating one's own thinking.

"We all participate in society and have opinions about society and how it works and how it's not working," says Greg Miller, a freelance science journalist based in Portland, Oregon. "I think reporting in social sciences you have to be aware of your own biases and be on the watch for confirmation bias."

For example, think about all those mental health stories that came out of the pandemic. Many presented a seemingly universal take—there will be a rise in postpandemic anxiety in the general public, for example—only to overlook entire population groups. Will essential workers who have been out and about all this time experience the same sort of anxiety? What about parents whose children have been attending school in person the whole time? What about tight-knit immigrant communities wishing to reunite? Ask yourself questions, such as: How do I feel about this issue? Did this idea come from my very like-minded Twitter audience? How can I find out if others outside my orbit feel the same way?

The best approach to tackling complex stories is to remain humble, says Justin Fox, a columnist at Bloomberg Opinion. The only pieces Fox regrets, he says, are those in which he "spoke with undue certainty."

Coming to grips with one's own perspective is often most critical when writing about issues of identity and race. "We are constantly having conversations about objectivity," says Leah Donnella, an editor at *Code Switch*, a National Public Radio podcast devoted to covering issues related to race.

Rather than striving for that neutral, authoritative voice common to classic journalism, the *Code Switch* team goes conversational, Donnella says. When the team scripts episodes, for instance, they ask speakers to evaluate whether the line is something they suspect or know. Both approaches are valid as long as those assumptions are also made clear to listeners, Donnella says. "We just try to be really clear about where people are coming from."

The Takeaway

At first glance, the paradox of choice theory would seem to imply that the social science journalist's job is impossible: Too much choice in what to cover (all of society) will lead to uncertainty and indecision. But a social science journalist needs to seek out stories that pare down the world to a series of tangible but essential questions: What have researchers asked their study participants? What have they not asked? Who is included? How is the study framed? How robust are those findings?

The social science journalist, in other words, holds the keys to illuminating some of society's greatest mysteries. (Oh, and that paradox of choice theory? It may, or may not, be reproducible. It's complicated.)

24

Science and Justice

Rod McCullom

Rod McCullom specializes in reporting on the intersections of science, technology, medicine, and society. His focus areas include artificial intelligence, biometrics, brain and cognitive sciences, gun violence, infectious disease, and the science of crime, trauma, and violence. He has reported from across sub-Saharan Africa and been awarded reporting fellowships to Australia, Ethiopia, Kenya, South Africa, Spain, and Zambia. Rod reports the Convictions *column on the science of justice for* Undark Magazine. *His column "Facial Recognition Is Both Biased and Understudied" was one of three finalists for the National Association of Science Writers' 2018 Science in Society Awards. He is a contributor to* Undark, Scientific American, Nature, *the* Atlantic, MIT Technology Review, *the* Nation, *and other publications. Rod attended the University of Chicago and was a 2016 Knight Science Journalism Fellow at the Massachusetts Institute of Technology.*

Reporting on crime, policing, the courts, incarceration, and criminal violence has long been the bread and butter of the news media, and it's easy to understand why. For starters, there's plenty of material to work with—particularly in the United States, where violent crimes like assault, homicide, robbery, and sexual assault occur about every 25 seconds. A property crime like auto theft or burglary happens about every 4 seconds. That is about how long it takes most people to read this sentence.

Not only do these issues involve profound personal stakes for the individuals involved, but also they inspire a curiosity that is, at its heart, very human. "If it bleeds, it leads," has been an informal motto in many newsrooms for over a century.

What's can be less obvious, however, is that that virtually all of us are impacted by the infrastructure of criminal justice—and that there are questions of science, both large and small, at its very core. Perhaps you've had your sleep disturbed by the sound of police sirens late at night or had your car's license plates quietly scanned by law enforcement—on the hunt for outstanding

Rod McCullom, *Science and Justice* In: *A Tactical Guide to Science Journalism.* Edited by: Deborah Blum, Ashley Smart, and Tom Zeller Jr., Oxford University Press. © Oxford University Press 2022. DOI: 10.1093/oso/9780197551509.003.0025 tickets and warrants—as you drive down the street. Or maybe you used your fingerprint to gain access to your phone or computer this morning?

Reporting on the science of justice includes scrutiny of the methods, processes, technologies, and interventions that comprise the criminal justice system: crime, policing, courts, mass incarceration, and parole and probation. Inside this broad portfolio sits a wide array of scientific assumption and technological innovation, from artificial intelligence (AI), biometrics, computer science, and epidemiology to forensics, genetics, gun violence, neuroscience, and psychology.

Recognition of these realities—and the need for journalistic investigation of them—has been growing. Ashley Southall, Eli Rosenberg, and Virginia Hughes of the *New York Times*, for example, have reported on the evolving role of familial DNA in solving crimes. *MIT Technology Review*'s Karen Hao has reported on the many complications surrounding facial recognition and AI-based policing technologies. And Aleszu Bajak has reported on policing technologies and big data for *Undark Magazine* and *USA Today*.

But, overall, there has been comparatively little journalistic scrutiny of the science—or lack of it—that underpins the criminal justice system. From the rise of algorithms that help judges make decisions about guilt or innocence to the advent of facial recognition technology used in policing (and the inherent biases baked into both), crime and punishment are no longer purely social constructs: They are scientific ones, too.

This means that readers will be well served by innovative and nuanced reporting that looks carefully at data used to justify decision-making in the criminal justice system, that scrutinizes new policing technologies, and that asks much-needed questions about how certain measures are deployed in the name of keeping the peace. And, at the same time, it is crucial to investigate the research and ask if these measure or technologies are even necessary.

Context Matters . . .

The best science writing, of course, goes beyond a simple cut and paste of a press release about a new research study. We have learned to ask basic questions about the research, such as data selection, study limitations, funding, and conflicts of interest. Outside experts will add depth and nuance to news coverage of almost any science story.

Reporting on the science of justice is no different. The reporting needs to go well beyond rewriting press releases by law enforcement, prosecutors, or their funded research partners. Unfortunately, that lack of broader context is often lacking or missing in many local news reports. It is crucial that science writers broaden their understanding of concepts like crime, violence, and justice, as well as the technologies used in law enforcement and incarceration.

At the moment, there is a national conversation—the Black Lives Matter movement being in the forefront—around policing and racial, social, and economic justice. It was easy for many people to reduce the protests to hashtags after the fatal police shootings of Michael Brown and Eric Garner in 2014. But since then, our understanding of these issues has evolved. Editors and reporters at the *Guardian* and the *Washington Post*, for example, created police homicide databases that demonstrated that African Americans and Native Americans are killed by police at much higher rates than Whites. And, for the first time, new research has documented that the Black Lives Matter protests have forced local police agencies to become more transparent, dramatically increased the use of police body cameras and, in some cities, even led to a decrease in police use of lethal force.

Science journalists need to investigate the problems that criminal justice interventions or technologies are intended to address. And, at the same time, they must continually investigate the research methodologies to identify limitations, flaws, and, crucially, bias. "It's important to see how the science is working," says Glenn Ellis, a bioethicist and research bioethics fellow at Harvard Medical School. "The public relies on science writers"—especially those covering issues of life, health, crime, and justice, he says: "to write stories so that we understand what to focus on and what is important."

So dig deep to understand how various interventions work and their performance metrics. Sometimes you will discover something surprising.

One example is acoustic gunshot detection technology. This technology uses GPS and machine learning software to "triangulate" gunfire location detected by dozens of audio sensors that many police departments install across both low-income and high-crime neighborhoods. The Chicago Police Department began expanding this technology across the city in recent years in response to soaring levels of gun violence. Police officials boasted that the technology was working, and shootings were "down."

As a Chicago resident, I saw these audio sensors installed on streetlights in my former neighborhood. I have reported extensively on gun violence and wanted to know how they worked and how they stopped shootings. For an early *Undark* column that was later picked up by the *Atlantic*, I interviewed several outside experts and learned that while the technology detects gunfire, there is no scientific evidence that it *reduces* gunfire. The reduction in gun violence in some neighborhoods could instead be due to gentrification, one expert noted. And, more importantly, most gun violence cases in the city were not solved, and the technology has not been demonstrated to help rush victims to trauma centers any faster.

. . . and Stay Skeptical

It is important for science writers to develop a healthy dose of skepticism of policing technology. Many science writers are too deferential to new technologies and fail to question their necessity, say Ellis and Nicol Turner Lee, an applied sociologist at the Brookings Institution whose research focuses on the digital divide.

Most science writers who report on policing technologies such as facial recognition, for example, "understand the technical nuances of the technologies but they may be less aware of the sociological implications on how these technologies impact certain populations," says Turner Lee. "There need to be more interdisciplinary conversations among scientists, social scientists, and ethicists" with science writers, she adds.

Science writers should not be afraid to ask researchers, technologists, and authorities to disclose the limitations of technologies and interventions, say Turner Lee and Ellis. "I would love for scientists to show some humility on the outcomes they are predicting and [offer] some suggestions around fallibility," adds Turner Lee.

Consider the reporting around AI-powered policing technologies. "Usually, the conversation is very myopic and one-sided, where the reporter frames the problem as: 'Violence is an issue. AI can solve it.'" says Desmond U. Patton, an associate professor of social work at Columbia University whose research explores the intersection of social work, technology, and AI.

Instead, he says, reporters should ask critical, pointed questions when covering any proposed AI technologies, such as: " 'At whose expense?' or 'Is this the right tool?' or 'Is it telling us everything we need to know?' " "Violence is an epidemic," he says. "To what extent does AI help or harm individuals who might be affected by violence?"

This lack of critical analysis has been echoed by numerous researchers and scientists throughout my reporting career. Keep this in mind when reporting on science and justice: Ask the questions that are not being asked by other science writers. Look at the reporting on your story and see how it could become more nuanced.

For example, in 2017, I began investigating racial bias in facial recognition technology used in law enforcement. This was around the time when issues of racial bias across AI were gradually being reported more in the news media.

Tom Zeller Jr., *Undark's* editor-in-chief and one of the editors of this guide, asked: "What could you do differently?"

We searched and found that the extent and underlying basis of racial bias in facial recognition was significantly understudied. At that time, only two or three relevant studies investigated or documented bias in facial recognition technologies that were scanning millions of faces across the country.

The lack of research became the focus of a column "Facial Recognition Is Both Biased and Understudied" that attracted the attention of researchers, academics, and policymakers. It was an early and crucial lesson for me in learning to survey the reporting landscape, to determine what is missing, and to fill the void with critical and nuanced reporting.

Incarceration and Race

Your reporting on science and justice will eventually lead you to stories on prisons, jails, and incarceration, especially if you are based in the United States. The United States is the global leader in mass incarceration: Despite having only about 5 percent of the world's population, we also have about one quarter of the worldwide incarcerated population.

As noted, context is key. African Americans comprise about 13 percent of the nation's population and about 38 percent of the population in state prisons. Black people are "incarcerated in state prisons at a rate of 5.1 times the imprisonment of whites," according to the Sentencing Project, a nonprofit research and advocacy group.

For decades now, researchers have documented racial bias in every area of the criminal justice system, including policing, courts, and corrections. This means that it should be difficult *not* to acknowledge racial bias when reporting on science, health, medicine, and mass incarceration. But acknowledgment of the proverbial "elephant in the room" does not always happen and becomes glaringly obvious to researchers and science writers—such as this one—who are African American.

Science writers are largely White and educated, and those privileges are apparent in their reporting on the science of crime and justice, Ellis and Patton note. This often means that disparities or race are minimized or go mentioned. "Science writers are products of society in that many see incarceration as a form of discarding trash" and inmates "as disposable," says Ellis.

"There are two million incarcerated people" right now in prisons and jails, Ellis adds, but they are rarely discussed in science. "So even if science writers are reporting on Covid-19, they're not looking at this population" in terms of vaccine distribution, for example, he adds.

Another recent example of how these privileges and disparities can impact news coverage is the limited reporting of data and science around the impact of mass incarceration on the wider community spread of Covid-19.

I was talking on the phone to a Chicago friend one evening early in the pandemic. He said that his cousin was recently released from Cook County Jail—one of the largest single-site jails in the United States—to relieve overcrowding. The cousin did not know that he was Covid-19 positive and inadvertently spread infections to several family members.

I had read previous reports on mass releases of inmates to mitigate epidemics inside correctional facilities, but hearing this personal anecdote made me realize that there was a national story around community spread and racial disparities because so many formerly incarcerated were returning to Black neighborhoods.

An underappreciated paper published early in the pandemic was the first scientific attempt to quantify the community spread from jail releases. My reporting for *Undark* wasn't the first to highlight this paper, but it helped to highlight the broader, national implications of mass incarceration's possible impact on Covid-19 racial disparities.

Personal Narratives

As in most areas of science journalism, stories that rely on the experiences of real people to illustrate the impact and extent of criminal justice science often lead to more compelling and engaging narratives. Don't rely solely on those in authority positions. Think also about including personal stories from family members, neighbors, victims, suspects, defendants, and perpetrators, along with researchers and scientists. They will only enrich your coverage.

Take, for example, reporting on the arrest of suspected serial killers. There has been much compelling science reporting in recent years around genealogy forensics, familial DNA technologies, and ancestry databases, all of which have been used to track down some of these suspects. In such cases, the family members of victims and suspected killers have become crucial characters in coverage of both the science and law enforcement sides of the story.

The Takeaway

Crime and justice impact everyone, but there is little skilled reporting on the underlying science. Find a niche. Do not be afraid to specialize and stray off the beaten path.

It is important to report on new research into policing technologies, but it is equally important to find out how it impacts society. Always question new technologies and ask: Who will it help and hurt? How much will this cost? Do we really need it?

Try to avoid stereotyping suspects or entire communities. Always look for personal stories, especially when reporting on abstract or complex science. And remember that race will always be a factor in reporting on crime and justice. Race and bias are always lurking in courtrooms and police investigations, says bioethicist Glenn Ellis: "It's like smoke. You know it's there, but you might not be able to see it."

25
Physics

Ashley Smart

*Ashley Smart is the associate director of the Knight Science Journalism Program
at the Massachusetts Institute of Technology and a senior editor at* Undark *magazine. Prior to that, he spent 8 years as an editor and reporter at* Physics Today
magazine. His writing has also appeared in Quanta, Chemical & Engineering
News, Wired, *and other publications. Smart was a 2015–2016 Knight Science
Journalism Fellow and holds a PhD in chemical and biological engineering from
Northwestern University.*

In the summer of 2008, a few miles outside of Geneva, Switzerland, physicists
began revving up the largest subatomic particle smasher ever built, the Large
Hadron Collider of the European Organization for Nuclear Research (CERN).
Known as the LHC, the nearly $5 billion experiment represented physics'
best hope of observing the Higgs boson, the final missing puzzle piece in the
standard model of physics. Its long-awaited grand opening was heralded with
headlines in news outlets around the world.

Chris Wilson, then a contributor to *Slate* magazine, took it as an opportune
moment to level a broadside against the small community of journalists who
cover physics. He surveyed the at-times-colorful press coverage of the LHC,
measured it against the more matter-of-fact science communication offered
up by the physicists themselves, and delivered a stinging judgment: "On the
whole, the best writing about physics for a general audience seems to come
from physicists, not journalists."

It is a point on which reasonable people could disagree. Physics, maybe
more than any other scientific discipline, is replete with practitioners who
excel at writing about their craft. I myself was inspired to write about science
in part by the elegant prose of the likes of Richard Feynman, Carl Sagan, and
Brian Greene, and I am equally impressed by today's younger, more diverse,
crop of physicist writers.

But the question of who writes best, physicists or physics journalists, is in
many ways beside the point. We do different jobs.

Ashley Smart, *Physics* In: *A Tactical Guide to Science Journalism.* Edited by: Deborah Blum, Ashley Smart, and Tom Zeller Jr.,
Oxford University Press. © Oxford University Press 2022. DOI: 10.1093/oso/9780197551509.003.0026

Like physicists, we physics reporters strive to capture the wonder and weirdness of a pursuit that continually pushes the limits of human understanding. And we try our best to explain it in ways that will be clear and meaningful to the layperson. But, crucially, we write from a vantage point of having little, if any, skin in the game. We have no pet theories to promote, no funding agencies to please. And so we are privileged to be able to find and tell stories that the physicists, by virtue of their station, cannot.

This challenge of reporting, as an outsider, on one of science's most notoriously difficult disciplines is not an easy one. If you are considering it, know that you will continually encounter concepts that lie frustratingly beyond your ken. To wrestle them within reach, you will need to grapple with equations and esoteric thought experiments. You will need to ask very stupid questions to very smart people. And if you are like many physics reporters I know, it will take you years to truly find your footing.

But if you stick with it—if you can develop a nose for good stories, report them with savvy, and tell them with grace—your reward will be handsome. You'll have a passenger seat on a journey to unravel the most fundamental mysteries of the universe we inhabit—and maybe even some we don't.

First, you'll just need to find a story to tell.

Know the Big Questions

The physics beat is vaster than you think.

Sure, it has its bread-and-butter subjects. You'll want to get to know the oddities and quirks of quantum mechanics; the fiery physics of stars, nebulae, and black holes; and the zoo of fundamental particles that make up atoms, molecules, and all matter.

But noodle around a bit, and you'll find there's an entire world of interesting topics beyond that horizon. "Physics asks some of the biggest questions in science," says *Physics Today* editor-in-chief Charles Day. "But it also asks practical questions: How can the efficiency of solar cells be improved? How robust is the electrical grid?"

Day suggests that, as the hordes "zig" toward stories about particle physics and astrophysics, more venturesome physics reporters may find success "zagging" toward undercovered fields like acoustics, biophysics, and condensed matter. Get to know the statistical physicists who are trying to understand how proteins fold; the atomic physicists using lasers to build ultraprecise clocks; and the econophysicists trying to tease out patterns in financial markets. As a former colleague of mine used to say: *Everything is physics if you look closely enough.*

If you decide to venture into these less charted waters—and you should—you'll need to develop your own instincts for what constitutes a good story. Try not to take your cues from attention-grabbing press releases. To find stories that will endure, Day suggests, pay less attention to the flashiness of a study's results and more to the importance of the question it addresses. "If the question itself isn't important," he explains, "the answer likely won't be either."

To know what those important questions are, you'll have to talk to physicists—lots of them. One tried-and-tested approach is to mill about at conferences. The American Physical Society hosts two of the largest, in March and April, as well as numerous smaller specialists' conferences. I like to look for presentation rooms bulging with standing-room-only crowds, then elbow my way inside to see what the fuss is about.

If you live near a university with a physics department, drop in for the occasional department colloquium. Reach out to professors who are doing work that interests you and try to arrange to visit their labs. You may not find a scoop, but you'll become familiar with the important questions that they and other physicists are trying to answer. And that familiarity will prepare you to recognize a scoop when one does come along.

That's the position *Quanta* editor and senior writer Natalie Wolchover found herself in the summer of 2016, when two research teams from the LHC were preparing to present their latest results at a conference in Chicago. Four years earlier, the same teams had set the science world abuzz with the discovery of the Higgs boson. Now, the collider had ramped up to full throttle, and it was smashing particles at energies even higher than those that revealed the Higgs. Researchers hoped the debris from these more violent collisions would reveal never-before-seen particles—perhaps ingredients of dark matter or some as-yet-unknown sibling of the Higgs. "It was sort of like the moment when the Large Hadron Collider was gonna show its cards," Wolchover recalls.

But as the teams presented their new data in Chicago, it became clear that they had found, well, nothing.

Wolchover recognized that what may have seemed like a disappointing result actually had profound implications for some of the biggest questions in physics. It dealt a blow, for instance, to a popular but controversial particle physics theory known as supersymmetry. And it gave a boost to speculative theories about multiverses. Wolchover wrote about those and other implications in what would go on to become an award-winning feature story.

"I think that was just an example where the news peg, if you just heard about it in a vacuum, you wouldn't know why it was a big deal," Wolchover said. Indeed, a lesser journalist, fixated on the disappointment of the null result, might have missed the story entirely—and many of us did.

Get Up to Speed, With Speed

You have your story idea. Will you be able to make sense of the science and decipher what it all means? Will you be able to distill concepts to their essence and understand their caveats? On a tight deadline, or even a loose one, can you *get up to speed*?

Those questions haunt every science writer, but for the physics reporter they loom especially large. So much of physics defies human intuition. We expect that an atom fired toward a pair of slits in a wall will travel through *either* this slit *or* that one, but quantum mechanics tells us that the reality can be more complicated. We are accustomed to seeing the swirling eddies in a fluid gradually die out, but superfluidity turns that expectation on its head.

Baffled by it all, a reporter might be tempted to throw her hands in the air, take the experts at their word and simply transcribe and translate to the best of her ability. That would be a mistake. Part of the job of any journalist is to exercise judgment—not only to blindly relay what sources tell them, but also to convince themselves that the story adds up. That doesn't change just because you're writing about one of the most mystifying disciplines known to humankind.

What this means is that if you're coming into physics reporting as I did, without an advanced degree in physics, the work of getting up to speed must begin well before you take on an assignment. When I first joined *Physics Today*, I bought all three volumes of *The Feynman Lectures on Physics* and read them cover to cover. The classic texts couldn't bring me up to date with state-of-the-art physics, but they provided an elegant, accessible overview that helped me begin to think more like a physicist. Review articles helped me push further into specific topics and subfields. Whatever your level, find an entry point that works for you and then go deeper in the directions that interest you most. Then, when you do need to get up to speed for an assignment, you'll have a running start.

When it comes time to drill down on a specific story, lean on experts to help you wrap your head around a paper's key findings, their significance, their caveats and limitations, and how they fit in with the bigger research picture. The more difficult the science, the more helpful it can be to talk with a range of experts who have different perspectives and different communication styles.

"You may interview a few people and not really 'get' it," says Wolchover. "And then you'll find someone who's really good at explaining it—or their explanation just for whatever reason works better for you." After talking with enough people, she says, "you get to a point where you can see the connections between what they're saying, and you realize that you're kind of figuring out the story by connecting their different tellings of it."

Understanding the connections between different tellings of the same story is especially important when you're reporting on an issue on which the physicists themselves disagree. One such story landed on my desk in 2018, when I received a tip about a paper on a long-running debate over supercooled water—water maintained in its liquid state below its normal freezing point. For the better part of a decade, two groups, one at the University of California, Berkeley, and another at Princeton, had been running nearly identical simulations of this peculiar state of water and getting results that were at odds with one another. Now, the Princeton team claimed to have uncovered a flaw in the Berkeley code, which the Berkeley researchers flatly denied.

One way to tell the story would have been simply to present both sides. But as I talked with different experts—and wrestled with the research literature—it became clear that the community of scientists who study supercooled water was coalescing around the Princeton group's argument. So in my story, I did my best to present both teams' cases, but I gave more weight to the Princeton team's case, and I tried to make clear to the reader the scientific justification for doing so. As a result, a story that could have come across as a mere tit-for-tat disagreement between two groups ended up being a much richer tale about the at-times-fitful march toward scientific consensus. To do the story justice, I had to cast a wide net for sources, I had to be willing to get into the technical weeds, and I had to be prepared to get up to speed, with speed.

Delight in the Details

Perhaps more than any other field, physics can seem foreign to our everyday experience. The action often unfolds over distances that are either inconceivably small or unimaginably large—distances measured, perhaps, in billionths of meters or millions of light years. The phenomena play out in mind-bogglingly brief instances—picoseconds, attoseconds—or over epochs that are inhumanly long. As a result, when it comes time to put pen to paper, the physics reporter faces a special kind of dilemma.

"The challenge," says Anil Ananthaswamy, former physics editor at *New Scientist* and author of several critically acclaimed books about theoretical physics, "is that the concepts you're dealing with are pretty difficult to wrap your head around, and you want to somehow draw people in."

A well-placed analogy to everyday life can often help bring complex ideas into focus for the lay reader. But beware the temptation to sensationalize and embellish—to strain metaphors, lean in to hyperbole, or, worse, deploy gratu-itous references to Albert Einstein. Such embellishments can distract readers from the underlying science, says Ananthaswamy. For him, a more effective way to draw the reader in to the abstract concepts of physics is to anchor those concepts in concrete, physical details.

A case in point: You will never *see* neutrinos, the tiny subatomic particles so difficult to detect that they've been dubbed "ghost particles." You will never *feel* the heat of the nuclear furnace, deep in the Sun's core, where most of our solar system's neutrinos are generated. And you'll never *hear* neutrinos, though trillions of them will whiz right through your body during the time it takes you to read this sentence.

But it is very much possible to see, feel, and hear Siberia's Lake Baikal, where for nearly two decades an underwater "neutrino telescope" has been detecting the elusive particles. So when Ananthaswamy set out to write about neutrinos for his 2010 book *The Edge of Physics*, he made the perilous trip out onto the frozen surface of Lake Baikal to see the telescope for himself. In his subse-quent book, he described the details of the travel, the geography, the place, and the people he encountered. He wrote about the scientific reasons for locating the experiment in such a frigid, bleak landscape. And then, slowly, he brought in the deep, theoretical questions about neutrinos and the physics of detecting them.

"Physics, no matter how theoretical it is, is about the physical world," Ananthaswamy explains. "It's much easier to explain what dark energy is when you actually talk about the experiments that discovered the presence of dark energy," he says. "The concept [of quantum mechanics' double-slit ex-periment] can feel very esoteric, but this experiment has been done over and over again—and you can use these physical details." By skillfully deploying these physical details, you may find it is possible to make physics' puzzling abstractions a little less, well, abstract.

The Takeaway

In his 2008 *Slate* article, Wilson wrote that one reason he preferred the writing of physicists over that of physics reporters is because physicists "don't try so hard to make you care," that they don't condescend to the reader with colorful prose and melodrama.

There's a grain of truth there. It can seem difficult to convince the average Jane or Joe why they should care about the very fundamental questions that physicists tend to ask about the world. Some physics research has obvious practical implications for society, but, frankly, most doesn't. In the struggle to make physics seem relatable, it's easy for a reporter to overdo it.

But never underestimate the capacity of average Jane and average Joe to be curious for curiosity's sake—to appreciate the wonder of a world that is mysterious beyond anyone's wildest imagination and to marvel at the quintessentially human quest to decipher it. If you can convey that sense of wonder, if you can satisfy even an ounce of that curiosity—and if you can do so with reporting that is accurate and authentic—that may well be all the relatability you need.

26
Genetics

Antonio Regalado

Antonio Regalado is the biomedicine reporter at MIT Technology Review, *where he covers genetic engineering. He loves a good scoop about scientists modifying DNA—even by other people. Prior to joining* Technology Review *in 2011, he was a correspondent for* Science *magazine in Brazil and was the science reporter at the* Wall Street Journal *from 2000 to 2006. He appeared in the CRISPR documentary* Human Nature *and has won the Hastings Center Awards for Excellence in Journalism on Ethics and Reprogenetics and the National Institute for Health Care Management 2021 trade award for stories about Covid-19. He lives in Cambridge, Massachusetts.*

My mouse hovered anxiously over the "publish" button before I made one of most consequential clicks ever, for me and certainly for He Jiankui, the young Chinese scientist whose secretive project to create gene-edited children I was about to make public.

That November 2018 story in *MIT Technology Review*, "Chinese Scientists Are Creating CRISPR Babies," was based on documents I'd hit upon online that described He's creation of the first genetically modified (GM) humans resistant to HIV. He and his public relations (PR) team wouldn't confirm anything, so that's why I clicked with some trepidation. Had they really gone ahead and used a gene-editing tool, called CRISPR, to genetically modify humans, despite calls by scientific societies not to do it, at least not yet? Within minutes the story was being read by hundreds of thousands of people thanks to a swirling red alarm bell on Drudgereport.com.

This story was breaking news, but at the same time old and familiar. It was really about "playing God" and how the advance of science can collide with social taboos. Dr. He would soon be compared to Dr. Frankenstein, although comparisons to Prometheus stealing fire from Mt. Olympus might have been apt. And like Prometheus, he'd pay the price. His experiment was sloppily done and widely condemned as premature.

Antonio Regalado, *Genetics* In: *A Tactical Guide to Science Journalism.* Edited by: Deborah Blum, Ashley Smart, and Tom Zeller Jr., Oxford University Press. © Oxford University Press 2022. DOI: 10.1093/oso/9780197551509.003.0027

He had told friends that he believed gene-modified people were inevitable and someone needed to go first. Confidants told him it would take a scientific "cowboy" to bend the rules and deliver the big breakthrough.

And you know what? I think there is a little cowboy in all of us when it comes to DNA. We order away for 23andMe tests ready to discover family secrets, and we tumble into bed together to create new people, hoping for the best. And who can resist news stories like "One Sperm Donor. 36 Children. A Mess of Lawsuits" or "Biohacker Regrets Publicly Injecting Himself with CRISPR." Sarah Zhang, the journalist who wrote those stories for the *Atlantic*, told me she thinks nothing captures the imagination more than genetics, which after all explains who we are and how we got this way. Sarah reminded me there's even a phrase for this feeling: "It's in my DNA."

Is it in yours? The first step to cover genetics well is the easy part: falling in love with the subject matter. If you do, late nights poring over *Nature* papers and following footnotes will be a thrill. When I polled my followers on social media—journalists and genetics enthusiasts mostly—several said they'd gotten hooked by watching the 1997 sci-noir film *Gattaca*, about a world divided into DNA elites and have nots. Others mentioned *Jurassic Park*. Some can't put down Richard Dawkins' book *The Selfish Gene*, which imagines a scrum of genes all competing for their individual survival. You and I? We're just convenient ways for them to carry on.

Making the case to your editor that DNA is important should be easy—after all, it's the instruction set for every living thing on the planet. What's more, there's a solid argument that genetics will be the primary scientific force shaping the next 100 years. "If the 20th century was the century of physics, the 21st century will be the century of biology," the geneticists J. Craig Venter and Daniel Cohen wrote in 2003, shortly after the sequencing of the human genome. "While combustion, electricity and nuclear power defined scientific advance in the last century, the new biology of genome research . . . will define the next."

We're shin deep in the century now, and gene science is crackling through industries and across the front pages. Eight of the top 10 drugs sold in America are proteins or antibodies manufactured in vats from synthetic human genes. More than 30 million people have signed up for consumer genetics tests. As I write this, with the world entering the second year of the Covid-19 pandemic, everyone is glued to breaking news about the newest mutations to the germ's genome. Did I mention reporters can make an entire career writing about genetics?

Finding Your Stories

So where does genetics news come from? A sociologist of science I know, Benjamin Hurlbut, describes a model in which genetics advances from "science to technology to society." That's a framework that describes how gee-whiz discoveries make their way out of the lab, into useful applications, which then trigger societal implications—real consequences for real people. As a writer for a technology magazine, I like to cover the arrival of new, world-changing applications of DNA science. Here are a few that reporters on the beat follow closely:

> *Gene databases*: These are getting enormous. 23andMe alone has 10 million profiles. One exciting application is "genetic genealogy" to catch serial killers and solve cold cases.
>
> *Genome sequencing*: Genome sequencing is faster, cheaper, and more essential than ever. During the pandemic, speedy sequencing of the coronavirus kicked off the vaccine race.
>
> *Synthetic genomics*: Synthetic genomics have the ability to re-create a virus, and maybe one day a cell or animal, from mail-order DNA parts. This is the field that, in part, led to the controversial lab-leak theory of Covid-19 origins.
>
> *Gene therapy and gene editing*: Doctors are curing serious rare diseases by fixing or replacing genes. In the United States, for instance, CRISPR gene editing is treating African Americans with sickle cell disease.
>
> *mRNA vaccines*: The shots that are saving us from Covid-19 are mRNA vaccines. They directly add genetic instructions (as RNA) to people's cells, which help trigger immune defenses against the virus itself After vaccines, the trick might become a cheap-and-easy way to treat cancer and other diseases.

Searching for a Human Angle

An editor once told me that there are only two types of medical science stories: One is ego contra mundum, Latin for "myself against the world." That's the story of the iconoclastic scientist who sets out to make a discovery or prove something against all the odds and prevailing opinion (He Jiankui fell into this category). The other type is what I call the "crying mother" story. That's any story with a patient battling a terrible disease, a family trying to save a sick child, or otherwise focusing on human health ordeals. I call it that because an editor of mine, hoping to improve a genetics story that lacked drama, once asked me: "Can't you get a crying mother in there?"

Finding dramatic human stories in genetics has become easier than ever. People with rare genetic diseases are crowding by the hundreds on GoFundMe sites. Once a reporter would have struggled to find such stories—now we struggle to pick the right one to cover. One home run story in this vein was "One of a Kind," a 6,000-word narrative by Seth Mnookin that appeared in the *New Yorker* in July 2014. It was about Matt Might, a father on a quest to understand the brain disease affecting his young son. As it turned out, by sequencing his son's entire genome, doctors were able to pinpoint the genetic cause of his disease. The twist: The boy appeared to be the only person on the planet with this particular disorder.

Mnookin's story was effective because of how it used an extreme outlier case to pinpoint the latest amazing capabilities of gene science. What's more, armed with his son's genetic code, Might had started a heroic effort to find a cure. That meant he embodies both archetypal character types: He was both a parent in tears and a self-appointed scientist in an against-the-odds search for answers.

The lesson here is how patients and their families—particularly parents of kids with rare genetic diseases—can be a key resource and immediate portal into the big scientific issues of the day. Often, families are out there, on the web, or in Facebook groups, raising funds, seeking help, telling their stories, though usually for a much smaller audience than a journalist can bring to a topic. (Mnookin found out about Might's saga from a blog post that Might himself had written.) Working with sources who are already public about their medical problems also alleviates privacy considerations.

Scientists as Explorers

Not every genetics story has a patient in it. Sometimes, reporters will want to cover developments in the lab where the protagonists are professional scientists and the action takes place in a petri dish—or on a computer screen. That might seem boring at first, but not if you believe, as I do, that molecular biologists are the great explorers of our day. After all, you can't be the first person to fly across the ocean any longer or the first to plant a flag on the Antarctic pole. Instead, the great expeditions are into the invisible world of molecules. Think of Amelia Earhart in a white coat or Ponce de León with a PhD.

Precisely because molecular biology is invisible—no sled dogs, no frost-bitten fingers—a reporter needs to look for other ways to generate atmosphere and find drama. To get the reader in the right frame of mind, I sometimes use a writing tactic I call "crossing the threshold." Basically, this just means starting a story with as evocative an entry into the scientific world as you can come up with. I used this trick in a story I wrote for *Technology Review* about gene drives, a way of engineering animals to spread a gene that causes them to die off. It is being eyed as a way to rid tropical regions of mosquitoes spreading malaria, though it is controversial for its potential risks, like wiping out entire species. This story, called "The Extinction Invention," used an actual doorway as the threshold to introduce the gene drive innovation:

"A student led me through a steel door, under a powerful gust of air, and into a humid room heated to 83°F. Behind glass, mosquitoes clung to the sides of small cages covered in white netting. A warning sign read, "THIS CUBICLE HOUSES GENE DRIVE GM MOSQUITOES." It went on to caution that the insects' DNA contains a genetic element that has "a capacity to spread" at a "disproportionately high" rate.

The story later noted that the young researcher concocting these potentially ecosystem-altering organisms was a 27-year-old student named Andrew Hammond. "There are so many cool ways to build these," Hammond exulted. "There are so many easy things to do." I intended, of course, to contrast his youth and ambition with the weighty question of whether we are wise enough to engineer nature at all. Once again, are we playing God?

Finding the Controversy

In reporting on genetic technology, questions often arise about how scary, or how negative, the reporter's portrayal is going to be. For instance, is using DNA databases to catch serial killers justice served, or is it a violation of the privacy of the innocent people whose genes investigators are rifling through? Is engineering human embryos a major advance in healthcare, or is it ambitious scientists meddling with nature?

Controversy makes stories interesting, and it's important to raise tough questions, yet keep in mind these can turn into moral panics that don't last and may be rooted in superstitions we don't even appreciate we have. The novel *Frankenstein* was inspired by medical grave robbers who needed corpses to dissect, something that was forbidden in 19th-century Britain. But today, donating a body to science doesn't raise any alarm bells at all. So maybe in 2050 no one will be surprised that doctors use gene editing to improve a human embryo—just like in the movie *Gattaca*.

The Takeaway

Genetics is the story of technologies that are going to shape life in the 21st century—it's about mastery of the molecular world and, sooner than you think, of our own chemical makeup. All that creates a big, exciting, and important beat for a journalist: to tell the wider world about the latest developments in genetics and what that portends, both good and bad, for all of us.

To cover genetics well, you want to dive deep into the science, learn about new discoveries and technology, and then find a person, idea, or situation that makes the future impact clear to the reader. It's important for the public to be aware and get involved early; otherwise, scientists will simply move ahead based on their own motivations, like curiosity, winning grants, or starting companies.

The story of the CRISPR babies in China—the first GM humans—brought together many of these themes. Here was a young scientist, working in secrecy, and doing the inconceivable, or maybe the inevitable. Genetic journalists did an excellent job of investigating, including learning just how many top researchers in the United States knew of He's plans ahead of time but said nothing. The basic problem was that scientists were rushing to engineer our species without getting the approval of the rest of us first.

He Jiankui is now in prison serving a 3-year sentence for practicing medicine without a license. He had planned a big PR coup, but it all went wrong when reporters got the news out first. For now, germline editing is on hold. That doesn't mean we're not on a slippery slope to designer humans. We probably are. But now the public is aware of what genetic research is capable of creating. They're better informed, and as a result, scientists are more cautious about running ahead before society can get used to the idea of genetic engineering and establish some rules of the road for ethical research.

27

Technology

Megan Molteni

*Megan Molteni is a science and health reporter at STAT, covering genomic med-
icine, neuroscience, and reproductive tech. She joined STAT after 4 years as
a staff writer at* Wired, *where she wrote about artificial intelligence (AI), the
DNA testing industry, and genetic privacy before leading* Wired's *science coverage of the Covid-19 pandemic. She has broken stories about setbacks in the
gene-edited livestock industry and the expansion of invasive new DNA technologies into U.S. law enforcement. Her work has also appeared in* Popular Science,
Discover, Undark, *and* Aeon.

In the 1980s, being a tech reporter meant being a part of a small, specialized
kind of geek club. The first personal computers were just hitting the market,
and the modern internet was still a twinkle in Tim Berners-Lee's eye. As author
and *Wired* editor-at-large Steven Levy recalls it, all the technology journalists
writing for all the publications in New York could fit around one 20-person
table—which they did, often, to talk about microchips and floppy disks and
hackers. Today, all the technology reporters from the *New York Times* alone
would have to squeeze to fit.

"What happened is technology became the subject that became the subject
of everything," says Levy.

That is, in today's digitally interconnected world, every story is a technology
story. And every technology story is also a story about people and politics and
the economy and the environment. So while the tech beat is bigger and more
competitive than ever, there are nearly endless opportunities to find a way in.

Play the Anthropologist

Silicon Valley is an actual place—a low-lying stretch of silty soil once covered in plum orchards at the southern edge of the San Francisco Bay. And
many of the tech titans—Facebook, Apple, Alphabet, HP, Adobe, Intel, and

Megan Molteni, *Technology* In: *A Tactical Guide to Science Journalism.* Edited by: Deborah Blum, Ashley Smart, and
Tom Zeller Jr., Oxford University Press. © Oxford University Press 2022. DOI: 10.1093/oso/9780197551509.003.0028

Oracle—are still headquartered there. But as technology has become more diffuse, and the industry's influence widened, so too has *the idea* of Silicon Valley expanded beyond its physical boundaries.

These days, Silicon Valley is perhaps more than anything a stand-in for a particular way of thinking, dressing, and talking. As the *Guardian*'s Julia Carrie-Wong put it, if Wall Street is capitalism unvarnished, "Silicon Valley is capitalism euphemized."

Speaking Silicon Valley-ese is a useful skill for any technology reporter. It helps you understand the mindset of the people you want to talk to. And it makes it easier for them to trust you. "Every big tech company has its own language, its own mores, its own culture," says Levy, who has written detailed insider histories of Apple, Google, and most recently, Facebook. "It can be like visiting an indigenous tribe. It's very hard for outsiders to really get in. But if you spend a lot of time there and people see you know the language, then they get more comfortable and they tell you more."

The same lessons apply when speaking to the scientists who are pushing forward the fields of research that underlie many of today's advances in technology: neural networks, quantum computing, and synthetic DNA, to name a few.

Part of the job of the tech journalist is to pay attention to the places where cutting-edge science is going commercial, and another part is evaluating how well those products work once they hit the markets. Developing deep sourcing within relevant academic fields is a critical component for being able to do both of those well. And scientific fields, like tech companies, also have their own lingo and jargon.

You don't have to get a PhD, but identifying the journals where exciting science is happening and regularly reading the abstracts of new papers can give you both a sense of where powerful new tools are emerging and how scientists are talking about them. I've found that researchers are much more willing to spend time with you if you can demonstrate at least some fluency in the thing they're studying. And the best stories come from being submerged in these different worlds so you can identify when a big change is happening within them before that innovation breaks into the mainstream.

Don't Trust the Demo

On December 9, 1968, in front of about 1,000 people seated inside San Francisco's Civic Auditorium, a Stanford engineer named Doug Engelbart showed off a wooden mouse with which he could manipulate electronic text on a computer. He also popped open programs in different screens and virtually collaborated with colleagues miles away. When it was all over, the room burst into standing applause.

"It was the mother of all demos," Levy wrote in *Insanely Great*, his 1994 book about one of the iconic products that would grow out of Engelbart's revolutionary presentation: Apple's Macintosh, the first family of personal computers.

Demos are one of those Silicon Valley traditions that hasn't changed a whole lot in the intervening decades. In essence, it's a demonstration, for a group of potential investors or curious journalists, of how a new product works. A well-placed story can legitimize a company's business, which could be helpful in securing funding or pumping up its value before going public. Or it could help paper over a spell of bad press. So it's important to view anything you see in a demo with a healthy dose of skepticism.

Take the self-driving car. For the last 10 years, Google has been strapping journalists into its driverless vehicles and ghost riding them around the sunny streets of Mountain View, California. "The first time you do that, it's shocking," says Cade Metz, a technology correspondent for the *New York Times* and author of the upcoming book *Genius Makers: The Mavericks Who Brought AI to Google, Facebook, and the World*. Going on that experience alone, it'd be easy believe the company's executives when they say these cars are going to be everywhere on the roads momentarily. And many reporters did, as the scores of ensuing stories predicting millions of self-driving cars would be on the road by 2020 can attest.

And yet, in 2021, truly self-driving cars aren't everywhere. Despite huge advances in AI and computer vision, going completely driver free has continued to elude engineers at not only Google's Waymo, but also Tesla, Argo AI, and others. The only place you'll find them is in specialized trials or beta-testing programs.

The takeaway, says Metz, is to remember that demos like a preprogrammed test drive are contained, tightly orchestrated spectacles. "They're designed to make you think a certain way," he says. "But it's hard to extrapolate from that."

To avoid falling into the demo hype trap, you have to do your homework. Track down any research papers the company has published or presented at a conference. Cultivate a circle of scientists who work in that field so that you can pass them what you find. Lean on them to provide independent assessments and necessary context. And if you can't find anything in the public sphere, that might be a red flag.

But another way, the one Metz prefers, is to skip the demos altogether and find stories by talking to people. One of his first reporting jobs was for the Register, a London-based site that applied the British tabloid sensibility to science and tech news. That all but guaranteed that most tech companies wouldn't talk to him. But it forced him to dig deeper to where the real story was. "It's so easy to take the stuff streaming out of these companies and there's often a lot of pressure to do it," says Metz. "What's harder is not to take it. But I think that's the way to do it."

Secrecy in the Valley

Silicon Valley was founded on open competition, creativity, and risk-taking. Because of an unusual feature of California law, any employee can quit at any time, taking their best ideas with them to start a rival company.

That culture hasn't changed much in the last half-century. But one thing that has changed is the rise of the public relations (PR) machine. Microsoft was the first to put PR people in the room with reporters. But the trend got supercharged during Steve Jobs's reign at Apple, where he demanded nuclear code–level classification around the release of new products, turning low-level decisions into state secrets and obfuscating with impunity to the wider public. That us-against-the-world mentality took hold, and today, the big tech companies have armies of narrative-shaping, reputation-managing PR professionals on their payrolls. Smaller companies have followed suit. Try to sit down with a new two-person startup, and you can bet there will be a PR flack in the room.

This trend has made tech companies increasingly tough to crack. But as much as spokespeople are obstacles to reporting, they are also a necessary part of it. Cultivating and maintaining positive relationships with PR people who pass you good information can pay off. Still, that doesn't mean you shouldn't hesitate to assert yourself when they're *not* being helpful.

As an example, you're in the room with the founders of a promising young startup, and you start asking some hard questions. The PR rep jumps in, interrupting the person you're trying to interview. Remind the PR rep that you're there to get the source's comments, not theirs. And that to do your job, you need to be able to have a conversation with them. If that doesn't work, and you're not getting what you need, you can always stop the interview. Remember, you don't work for them.

A bigger challenge lately is that many tech companies want to make their executives available only to speak on background as anonymous sources, to just "fill you in" and then ask to approve any on-the-record quotes, says Issie Lapowsky, a senior technology and politics reporter at Protocol. "I just don't think that's journalism," she says. Instead, if a company is insisting to speak on background, she'll agree to it, but on the condition that the source speak on the record at the end of the interview.

She doesn't love the compromise because the conversation invariably winds up getting duplicated and watered down. "But it's important for journalists to remember that doing that doesn't mean giving up all your editorial discretion," says Lapowsky. "You still have the power to differentiate between their honest answers and their PR answers and just because they gave you a quote doesn't mean you have to use it."

Another thing new tech reporters should know is that anyone visiting any of the big tech companies—and increasingly the smaller ones, too—will be asked at the front door to sign an NDA, a nondisclosure agreement. As a journalist, you are there to report. An NDA is antithetical to that mission because it gives the company power to decide what you can and can't say. And you have every right to decline to sign it.

Often, front desk staff aren't aware that this is an option, so they might make you wait in the lobby until your PR handlers come down and let you in. Some reporters sign things like "void" or "absolutely not" on the line, which would not legally bind you to the agreement. Provided your publication has a no-NDA policy that will back you up—and most should—you can always use that to diffuse the situation.

I have encountered a few situations, especially with biotechnology companies, where planned tours have been canceled when I've refused to sign an NDA at the door, which led to some less-than-ideal interviews in lobbies. But it's important to establish and hold firm to your rights to report on anything that's newsworthy that you see or hear.

There are, of course, other ways to get inside tech companies besides walking in the front door. Because of Silicon Valley's culture of mobility, tech workers hop between companies frequently. And it's easier to get them to open up about previous employers than current ones. LinkedIn is an invaluable tool for finding people like that, and the platform offers periodic trainings for journalists on how to use it for reporting. Those trainings also come with a LinkedIn Premium account, which provides you with a free way to message anyone. All the reporters I spoke to recommended it.

Following the antitech backlash in the wake of the 2016 election and the Facebook–Cambridge Analytica scandal, more tech workers have become emboldened to go on the record. But many won't for fear of breaking confidentiality agreements. That's why the best thing is to have documents, says Metz. "Emails, recordings, those are the best because everyone is protected but you know the information is real."

The Softer Side of Tech

So far, I've described what people traditionally think of when they think of tech journalism. Then there's what *MIT Tech Review* senior reporter Tanya Basu calls the "soft tech" beat, which she defines as stories for and about people who have no direct ties to Silicon Valley, but whose lives are nonetheless impacted by technology.

She's written about digital mourning in the time of Covid-19, online sleuths helping the FBI (Federal Bureau of Investigation) identify Capitol insurrectionists and how screen time is altering kids' brains. "I think especially with the pandemic, there's been a lot of loosening of these hard edges of what we think of as a tech story, because our lives have gotten so technology-reliant," says Basu.

The questions she winds up asking are more like what most people in the world want to know: How can I use technology to accomplish my goals? Or use it in an unexpected way to solve new problems? How is technology impacting my privacy? "What appeals to me is I can explore these subjects and be an idiot and ask questions about things that more seasoned tech reporters might just brush right past," says Basu.

As technology has become infused in every aspect of daily life, it has become a subject of study in its own right. Right now researchers are trying to understand how technology is changing the way individual humans learn, connect, form identities, build social networks, care for each other, work, live, create, procreate, and die. The internet, social media, facial recognition, AI, DNA tech, and genetic engineering are also operating on the population level—altering economies and politics and healthcare and the environment. These are all tech stories. And because researchers are tracking those changes, most of them can be science stories, too.

The Takeaway

Technology can be a daunting beat. You are trying to untangle hype from hope, to understand technically dense subjects, and to see around corners and inside the machinations of opaque organizations guarded by highly paid teams of PR professionals.

But most reporters got into journalism to speak truth to power, and no one is more powerful in our present moment than technology companies. In 2019,

Amazon, Alphabet/Google, Apple, and Facebook brought in a combined $773 billion in revenue—making them the 20th largest economy in the world. With billions of people using their platforms, services, and devices, these companies are operating at a scale never before achieved by any industry in human history. Their products and algorithms shape people's lives, often imperceptibly. It's our job to look into that black box and turn on the light.

28
Space

Nadia Drake

Nadia Drake is a contributing writer at National Geographic, *where she covers everything happening off-world (while sneaking into earthly jungles whenever she can get away from the cosmos). Her byline has recently appeared in* The New York Times, Scientific American, *and the* Atlantic. *Nadia has a PhD in genetics from Cornell University and a certificate in science communication from the University of California at Santa Cruz. She's written three* National Geographic *cover stories, and her work has won multiple awards from various scientific societies.*

Like space itself, the space beat is vast. It includes planetary science, cosmology, astronomy, astrophysics, policy, human space flight—really most things under and including the sun. As a space reporter, you might spend a few days working on a story about a strange exoplanet, then do a breaking news piece about a newly announced space mission, then interview a dozen astronauts for a feature about how seeing Earth from space changed their perspective on our watery little world.

Broadly, the same journalistic practices that produce a solid science story in any beat are applicable here. Cast a wide net when choosing sources, vet your information—whether human or published study—check your assumptions, seek outside opinions. Always get the name of the dog or the space robot.

But space does have some peculiar challenges: first, aliens. Humanity's obsession with extraterrestrial life is as bottomless as a black hole. Next, the chief U.S. government agency involved in space happenings is very popular and very concerned with its own image. As well, space is one of a few beats where you can write a story that has literally nothing to do with Earth—but everything we know about the universe starts here, with imperfect humans.

Last, space stories can often feel like distractions, especially during a pandemic, but the wonder they can inspire is priceless.

"I don't think wonder is a diversion, or a bit of glitter on the side of your serious vegetables journalism," says Ross Andersen, an editor at the *Atlantic*

Nadia Drake, *Space* In: *A Tactical Guide to Science Journalism*. Edited by: Deborah Blum, Ashley Smart, and Tom Zeller Jr., Oxford University Press. © Oxford University Press 2022. DOI: 10.1093/oso/9780197551509.003.0029 who reports on space stories. "We cover people who are doing some of the coolest things that any humans do or have done. That's fun! And you don't want to lose sight of that."

Be Wary of Aliens

Space—whether rovers exploring Mars, enigmatic cosmic explosions, or actual rocket science—attracts a lot of attention. Arguably, we have the luxury of covering a subject that many readers are inherently interested in. And that's great. But one of the reasons for that innate interest is: People really want to know if there's life beyond Earth. And they love reading about aliens—really, really love it.

"I think it's worth pausing and saying not just that people tend to be interested in aliens, but that people are interested in them for a good reason," Andersen says. "It's a super fascinating cosmic question—are we alone?—and the answer may be within the reach of science."

The search for extraterrestrial life forms is one of the most exciting scientific endeavors of our time, and there's no sense in censoring musings about aliens. But there are ways to approach the topic well and ways to do it badly. Let's start with the not so great. Teasing readers with a tabloidy headline about a mysterious radio signal—or interstellar asteroid—probably being the work of aliens is, in the vast majority of instances, doing everyone a disservice. It's a cheap tactic that ultimately will disappoint readers and erode trust, as will overhyping or inflating the importance of results.

Instead, if you're diving into a story about mysterious cosmic phenomena, don't be afraid of complexity. Explain why an attention-grabbing claim exists, walk readers through plausible alternatives, and introduce them to independent experts who can provide more context.

Here, as in all your stories, don't worry about the science being too hard to understand. Just explain it well. "Trust your reader to come with you," says Lisa Grossman, astronomy reporter at Science News. "They're reading this because they're already interested in it. Respect that."

For example, Andersen accidentally touched what he calls "the white hot stove" of people's interest in intelligent aliens in 2015, when he wrote a story about a star that was behaving in unexplainable ways. Called Tabby's Star (or, more cheekily, the WTF star), it sporadically dimmed, sometimes darkening to a smidge of its original brightness. Scientists wondered whether its weird fluctuations could be explained by massive, energy-harvesting structures in orbit.

Andersen's story about that speculation was a hit—and it explained why ETs (extraterrestrials) were even plausibly being considered. It carefully went through various explanations for the star's behavior, and it laid out a road map for solving the mystery. As a result of that work, Tabby's Star stayed in the public eye until "dust" turned out to be the rather mundane answer for its odd flickering, giving people a glimpse of how scientists solve puzzles.

"Astronomy is a gateway science, it gets kids excited. People love space," Grossman says. "And aliens are a gateway to conspiracy theories—but they're also a gateway to curiosity."

NASA Needs Its Own Chapter

NASA's (National Aeronautics and Space Administration's) press engine is a well-greased machine that functions extraordinarily well when promoting events or discoveries that the agency wants in the spotlight. Think slick visuals, dramatic animations, fact-intensive information kits, and spacecraft with Twitter accounts. Our job is to remain objective and not to get distracted from pursuing information that NASA may be less enthusiastic about sharing.

"They are, surprisingly, a very untransparent agency," says freelance journalist Sarah Scoles, who has written about the extreme stumbling blocks she encountered when submitting document requests under the Freedom of Information Act (FOIA).

Scoles's experience rings true on many levels, from submitting FOIAs to seeking outside comments on studies. The agency's various centers (there are 10 field centers scattered around the country, in addition to headquarters in Washington, D.C.) handle interview and information requests differently. Some are more restrictive than others.

In some instances, it helps to show up with solid information about why you're seeking an interview with a particular NASA employee, perhaps even citing their recent papers about a discovery or subfield you're covering. Another trick is to look for a researcher's alternate affiliation—many have secondary appointments at universities and can speak to you without representing NASA. Third, it's always good to develop relationships with public information officers and work them as you would any other source; fourth, be stubborn.

NASA is one of the few U.S. government agencies with a legitimate fan base. People buy NASA-branded merchandise, and the agency's main Instagram account has more than 63.5 million followers (roughly 30 times as many as the Department of the Interior's account, which posts gorgeous images of national parks, and a hundred times more than the National Oceanic and Atmospheric Administration and its feed full of Earth imagery). The agency even has its own TV station, where live broadcasts of launches and landings are immensely popular.

"There is, at least from the U.S. perspective, this fascination with NASA," says Alexandra Witze, a freelance journalist and correspondent for *Nature*. "NASA can do no wrong. It's maybe a little bit out of proportion to some of what the accomplishments of the agency actually are."

Keeping that perspective in mind will help you sift through and evaluate the mountain of information you'll encounter on this beat. And other space agencies exist. Japan, Russia, China, India, the United Arab Emirates, and the countries comprising the European Space Agency are incredibly active off-world. So are some private companies.

"Don't forget there's an entire world doing space exploration," Witze says. "Be aware of the context of space history, astronomical history, what we already know, who has already discovered what."

On the Subject of Preprints

Sometimes, science happens quickly and publicly, and it's our job to synthesize, scrutinize, and report on information as it becomes available.

In September 2020, astronomers announced that they'd found phosphine gas in the toxic clouds shrouding Venus, and that it could be the product of an aerial biosphere. Of course, more work was needed to rule out a nonbiological origin for the gas, but according to the study authors, they'd tried to do that and had failed.

It was a huge announcement. Almost immediately, scientists challenged the detection. Published in the journal *Nature Astronomy*, the observations (although peer reviewed) were not extremely convincing, and the team had performed some unconventional analytical methods to pull the phosphine signal from the planet's spectrum. To borrow a cliché, their extraordinary claim was not supported by extraordinary evidence.

Quickly, multiple teams independently processed the discovery data and concluded that the original analysis produced a spuriously strong signal. But unlike the original paper, those reports weren't peer reviewed—they initially appeared on the arXiv, an open-access preprint server that's been around since 1991 and is updated every weekday.

The arXiv is an essential tool for reporting on space discoveries. By scrolling through postings, you can get early scoops, find sources, and figure out who is on competing teams and just how competitive they are. Some posted papers have already been accepted for publication, while others are drafts of work still in progress—but each manuscript contains information specifying where it is in the review process.

"It has a really good reputation, both among astronomers and journalists," Witze says.

In the weeks following the phosphine announcement, the arXiv became the primary place for astronomers to debate the merits of a detection that had the potential to alter our conceptions about life in the cosmos. But many journalists who reported on those initial contradictory works were criticized for reporting on work that wasn't yet peer reviewed—even though that information was publicly available and highly relevant. In fact, as of press time, the debate is still ongoing.

Peer review is not a panacea. Plenty of wrong results have been published and been retracted. Lots of great work appears in preprints or happens in real time as astronomical discoveries are made. Our job is to report on what's happening: because that's science in action.

"It's a constant process of learning and iterating, and many, many findings turn out to be wrong, or at least not the whole truth, down the line," says Clara Moskowitz, an editor at *Scientific American.* "And that's not a failure—that's how science works.

Space Begins on Earth

The space beat might focus on everything happening off-world, but space begins here, on this planet.

Attempting to understand and explore our universe is a fundamentally human enterprise—and space sciences are vulnerable to the same flaws as the rest of humanity's endeavors. "Every space story starts on Earth and has something to do with Earth and is run by the people on Earth—unless there are alien space programs that I don't know about," Scoles says.

Consider your sources' motivations, funding, and biases. Be aware of social issues in the field. Like many areas of science, astronomy has undergone a very public reckoning over the last decade related to sexual harassment. Similarly, astronomers are working hard to promote diversity in the field and to fight systemic racism, sexism, and bigotry.

"Work to cover diversity issues but realize that it's fundamentally different than covering the science that people are doing," says freelance journalist Joshua Sokol. "It's harder for people, potentially. They're much more vulnerable."

Astronomy is also publicly reckoning with indigenous rights. Since 2014, astronomers have clashed with Native Hawaiians over building a mammoth, multi-billion-dollar telescope on Mauna Kea's summit. The proposed Thirty Meter Telescope would be Earth's biggest ground-based window to the cosmos—but for the mountain's protectors, it's one more instrument in a sacred space that already hosts too many telescopes.

Issues of land ownership, consent, and sacred spaces are perhaps more acute for astronomy than for some other beats because the profession's instruments often occupy the same mountaintops that have been sacred for millennia. And it's worth offering that context to readers.

"One thing I can confidently recommend is to pay attention to people outside of space—other academics, non-academics with other priorities, people from backgrounds that aren't well-represented within the space community," says Space.com's Meghan Bartels, who strives to include Indigenous issues in her work.

Human Space Flight

The space beat can feel like very low-stakes journalism (although arguably, we are covering the race to solve some of the highest stake questions around) except when it comes to human space flight.

"I remember being a young, new space reporter sent down to Cape Canaveral to cover space shuttle launches, learning that I needed to prewrite five versions of the launch story," Moskowitz says. In her draft folder, she had a successful launch story, a weather scrub story, a technical glitch scrub story, a "stray-boat-wandered-into-the-no-go-zone" (this actually happens) story—and the bad story, the story you never want to run. The one that makes your stomach twist as you write it because you're imagining dead astronauts.

"It was eerie and scary and really drove home the stakes," Moskowitz says.

Defying gravity and sending humans into orbit atop a column of rocket fire is risky. It's normal to be nervous when you're covering launches. But there are other things to keep in mind as you dive into this corner of the beat.

First, language matters, and it's important to eliminate sexist, colonialist framing from conversations about space exploration. For example, it's no longer acceptable to talk about "manned" space flight: Even NASA's style guide uses the word "crewed." Or, use "piloted" or "human space flight." And if you're writing about "colonizing" Mars, ask yourself if that's really what you mean. On Earth, colonialist expansion worked out quite poorly for Indigenous cultures, and maybe it's better to speak of humans simply "living" on Mars.

Second, private companies are major players in U.S. space flight. We've all heard of SpaceX and Blue Origin, but other companies are getting into the suborbital game, and still others are working on launch vehicles and payloads. Covering private companies is different from covering NASA.

"There are different levels of transparency and openness, and different expectations of transparency and openness," says Jeff Foust, senior writer at *SpaceNews*. "I know black holes exist because of the 'media at spacex.com' email address—your email goes in and never comes back out."

You can't FOIA private companies—and SpaceX is famous for ignoring reporters—but there are ways to get information from them. You might develop sources within the company. Spend some time reading (*really reading*) relevant appropriation bills and government contracts. Ask the Federal Aviation Administration about how SpaceX violated a launch license, rather than waiting for an answer from the company (or a late-night tweet from Elon Musk). And take a dive into required Securities and Exchange Commission filings.

"There are all sorts of interesting hidden nuggets in those documents that companies might not disclose in a press release, but which they are required to disclose to regulators," Foust says. "A lot of this is not what you would consider conventional space journalism. It's a hybrid with other beats, from policy to business, to technology as well."

The Takeaway

Science is a human endeavor, and it's crucial to remember that everything on the space beat is vulnerable to human foibles. Question motivations, scrutinize findings and funding, put discoveries in context, vet your sources, and avoid the one-stop "prophets of everything," as Sokol says. Watch your language. And don't lose your sense of wonder or ever stop wondering whether there's life beyond Earth.

29

Climate

Sarah Kaplan

Sarah Kaplan is a reporter at The Washington Post *covering humanity's response to climate change. Her job has taken her from the Arctic Circle to the top of the Empire State Building in search of solutions to the planet's most pressing problem. She joined the* Post *in 2014 as an intern in the Style section before working vampire hours on the overnight news team and covering Earth and the cosmos on the science desk. Sarah also contributes to the* Post's *national breaking news coverage, reporting on natural disasters, human violence, and the coronavirus pandemic. Her work has won awards from the American Geophysical Union and the American Association for the Advancement of Science and was included in the 2020 edition of* Best American Science and Nature Writing.

Climate change is transforming our home planet in abrupt and alarming ways. Scientists have spent decades confirming this truth. Public opinion polls show the vast majority of people recognize it. Countless people are already experiencing it firsthand.

The job of a climate journalist now is not to convince readers of reality, but to tell the story of how humanity confronts it.

That means we're not just communicating the science of climate change. We're holding powerful people accountable for the damage. We're raising questions about who and what gets valued in a world warped by prejudice and inequality. We're bearing witness to the loss of lives and livelihoods, homes and habitats.

I'll be honest: Sometimes it's terribly depressing.

But I still think it's the best job in journalism. Climate reporters have a chance to cover, and possibly shape, the biggest transformation of society since the Industrial Revolution. All the necessary ingredients for powerful storytelling are there: conflict, emotion, loss, and possibility. The subject can be as specific and intimate as the fate of a single animal, or it can be as broad as the future of the whole world.

Sarah Kaplan, *Climate* In: *A Tactical Guide to Science Journalism.* Edited by: Deborah Blum, Ashley Smart, and Tom Zeller Jr., Oxford University Press. © Oxford University Press 2022. DOI: 10.1093/oso/9780197551509.003.0030

Start With the Science

High up in almost every story I write is some variation on the following two sentences, based on findings from the United Nations' Intergovernmental Panel on Climate Change (IPCC): "Scientists say humanity must roughly halve its greenhouse gas emissions by 2030 to avoid the most catastrophic effects of warming. If global temperatures increase beyond 2°C (3.6°F) above the preindustrial average, the consequences could include total loss of coral reefs, destabilization of polar ice, large-scale sea-level rise and deadly weather extremes."

These basic scientific facts are the standards by which policy proposals get measured. They provide context for climate-related disasters. They are the stakes of our reporting—so we must learn them all.

Read the big studies, like the IPCC's assessment reports and the U.S. national climate assessments. They can be long and technical, but they also offer excellent summaries of the problem. You'll come away with an overview of how climate change is playing out in various communities and ecosystems, which will help you identify story targets. The contributor lists for these documents are also great places to find potential sources.

I also scan the major journals each week to see what's new. Even though I rarely cover single academic papers, reading the abstracts of papers can give me a sense of what questions the climate world is interested in and help me identify when the research is converging on a big issue.

Sometimes, a study can be the launch pad for a much larger project. A few years ago my colleagues on the *Washington Post* climate desk stumbled across research showing how Earth is warming unevenly. Realizing that the hottest spots could provide a hint of what a warmer future might look like, they conducted their own data analysis, identified places where average temperatures had already increased 2°C and reported 12 powerful stories about the effects of extreme climate change. In 2020 the "2C" series won a Pulitzer Prize for explanatory reporting.

Attending conferences is a great way to build relationships with researchers, find out what experiments are happening, and get yourself invited on fieldwork. Talking one on one gives you the opportunity to ask questions about things academics take for granted. Some of my favorite stories only happened because I followed up on a researcher's offhand comment that made me think: "Wait, what?"

Climate science is a massive field; not even Einstein could be expected to know everything about the chemistry of the greenhouse effect *and* the effects of a late frost on songbird migration patterns. I keep a roster of friendly researchers I know I can contact when I'm confused. They don't need to be the foremost experts on a subject; often, they're young PhD students I met on assignments or follow on Twitter. But they know a whole lot more than I do, and they can help me navigate nuanced data and tricky conclusions. Sometimes they'll even reach out with their own suggestions for stories, leading me to look into topics I might never have explored.

Make It Personal

I often think about stories as engines: complex but comprehensible machines that take readers someplace new. If science is the road that the story of climate change travels along, humanity is the fuel. People—their needs, their emotions, their choices—are what will give your journalism power.

No matter the assignment, start reporting by asking yourself: Who has something at stake here? Make a list. For a story about a proposed gas tax, that list might include scientists and economists, the legislator who introduced the bill, oil companies, car owners, electric vehicle manufacturers, public transit system operators, and people whose neighborhoods near highways are affected by pollution. Depending on the story (and the deadline), you might not interview everyone on your list. But this exercise will help you conceptualize who is driving the news, who will be affected by it, and what the practical and emotional consequences might be.

Profiles can be great vehicles for explaining scientific concepts; if readers become invested in a person, they will also become invested in whatever that person cares about. For a big feature about the changing Arctic, I wrote about the relationship between a veteran sea ice expert and a young scientist he mentored. Their conversations about the world he knew and the altered environment she would eventually share with her students gave emotional resonance to research that might otherwise feel remote to American readers.

Personal stories also help readers understand the tensions involved in tackling climate change. By exploring the motivations of an individual, or the tensions within a single family or community, journalists can make concepts like "decarbonization" and the "energy transition" concrete. One of my favorite pieces from the *Post*'s "2C" series was a story by Juliet Eilperin about the wrenching choices facing an Alaskan Native community where oil drilling has brought prosperity, but rising temperatures threaten a way of life.

The world needs these stories. Research shows that people's understanding, sense of urgency, and willingness to take action around climate change are influenced far more by emotional appeals than by recitations of facts.

But storytelling is not simply a ploy to motivate change. The emotional aspects of climate change are essential context for all the scientific, economic, and political issues it raises. Ways of life are being created and lost as the environment and our regulation of it changes. Survivors of increasingly frequent and extreme natural disasters are deciding whether to rebuild or to move on. Individuals and societies are being forced to consider what they value and what they are willing to do to protect it. Only by reporting these human stories, stories that are intimate as well as illuminating, can climate journalists fully inform our readers.

Go Where the Story Is

In environment stories, Earth itself is a character. Since it can't speak, you'll have to use your reporting and writing chops to bring it to life. I often utilize a trick I learned from *Tampa Bay Times* feature writer Lane DeGregory: Draw a vertical line down the center of each page in your notebook and use the left column to record what a source is telling you while filling the right side with everything else you observe. Consult each of your senses: How does the air smell? Does the frozen ground crunch underfoot? What color is the sunlight that streams through 1,000-year-old redwoods? Try to get as specific as possible and don't be afraid to ask your source to help you interpret what you see.

Spending time with sources at their homes, workplaces, and field sites will also give you a chance to look for the telling details and resonant instants that make them come to life. What photographs does the scientist have framed on her desk? How does the farmer react when he finds a row of plants ruined by floods? Often, the action you witness can communicate the emotions or tensions of a moment far more eloquently than a quote.

Some of the most difficult but rewarding field reporting I've done has been during natural disasters. Journalists must be judicious about how we describe the connection between individual weather events, such as hurricanes and wildfires, and larger shifts in climate. But science has become good enough that we can responsibly say how much warming has increased the risks of a given weather extreme. Stories told from these events on front lines of the climate crisis can bring its dangers home to readers in ways no academic study can achieve.

This work must be done responsibly. If the crisis isn't happening in your community, do your research before parachuting in. Learn the challenges people faced before the disaster, know which residents were already marginalized and had the least to lose. Make sure you have the equipment and skills you need to be safe; if an emergency medical services crew has to rescue you, you're taking resources from where they are needed. Center the story on the people who are most affected by the disaster and reflect their agency and resilience as well as their needs. Ask them: What do you want readers to know? What do you want officials to do?

Being on the scene when disaster strikes puts you in position to tell some really important stories. You will establish relationships with community members, who can become essential sources for longer term projects. You will see firsthand which systems fail and who gets left behind. Kendra Pierre-Louis, one of my climate journalism heroes, often says there's no such thing as a "natural" disaster—meaning the consequences of any event are invariably the result of human choices. Be on the lookout for these "unnatural" causes of suffering, and by the time you get back to your desk and start filing records requests, you'll have the makings of a great investigation already in your notebook.

But you won't always be able to report directly from the places you write about. The logistics may be too difficult; your newsroom might not have the funds. A deadly virus might make it impossible to leave your home, as has been the case for all of us working during the Covid-19 pandemic.

Even so, you can still report *for* a scene without being *on* the scene. Ask your sources to give you a video tour of their home, their lab, their church. If there's a specific event you're hoping to witness—a zoning board hearing, a funeral, the dissection of a diseased elk—see if your source will set up a phone or iPad in a corner and let you listen in.

In phone interviews, dig for particulars, even if it feels awkward. My friend Tracy Jan, who covers race and the economy for the *Post*, always asks what things smell like—not because she necessarily wants to include that detail in her story, but because it gets her sources thinking deeply about the kinds of vivid details she's seeking. And share something about yourself; if you bring your full humanity to the conversation, you will see more of your source's humanity in return.

Look for Hope

Last September, when one of the worst wildfire seasons on record was ravaging the West Coast, I came across an image of Earth taken from orbit showing North America swathed in smoke. The image reminded me of photographs captured by Apollo astronauts on the moon, and something about it made me burst into tears. It was so stark, seeing our precious blue marble literally on fire, thinking about the avoidable suffering of so many millions of people living underneath that haze.

Covering climate can make a person feel a bit like Cassandra from Greek myth—certain that catastrophe is coming, but unable to convince people to change. You will witness devastation and loss. You will cover countless insufficient efforts to confront the problem. You'll get bombarded by emails from companies that seem more eager to advertise their green credentials than actually address their emissions.

You'll deal with people who deny the reality of climate change, either because they don't trust the science or have a financial incentive to ignore it. This is a smaller group than you might think; research by the Yale Program on Climate Change Communication shows that large majorities of Americans are seriously concerned about the problem and think the government and industry should do more about it. To paraphrase journalist Emily Atkin, remember that it is not our task to convert people to care about climate (leave that to the activists). The job of journalists is to seek and share the truth, empowering people to become more engaged and effective citizens of the world.

It's vitally important to protect your mental health while doing this work; otherwise, the job will become unsustainable. Find activities that allow you to decompress and friends or a therapist who can help you process what you've seen. Take time to tune out from the news.

Make room for stories that illuminate the best of humanity. Find the teenager campaigning to get the county school bus fleet replaced by electric vehicles or the low-income neighborhood pioneering new ways to deal with urban heat. In addition to documenting the dangers of climate change, report on the possibility of a world where humanity has confronted it. People will only fight for a safer, fairer future if they believe such a future is possible. As a storyteller, you can help readers imagine what progress might look like, emboldening and empowering them to seek the transformation the planet demands.

Remember that Earth remains wondrous. There are endangered species that survive despite the odds, polluted watersheds that recover after decades of concerted community effort. There are snowstorms and songbirds and fireflies that glow in synchrony each summer. Octopuses can solve puzzles, and networks of fungi allow trees in a forest to communicate. Wombat poop forms a cube! The glorious diversity of life is as much a part of the environment beat as hurricanes and smokestacks.

Spend time in nature. Climb a mountain, walk on the beach. Watch robins forage for worms and grass seed in the park near your house. Keep your eyes open for strangeness and beauty. It will enrich your writing—but more importantly, it will enrich your life.

The Takeaway

Climate is a beast of a beat, replete with technical science, complicated politics, and sheer heartbreak. But these qualities are exactly what make it such a rich subject to cover. Your reporting could take you from the nation's top laboratories, to a remote wilderness, to the halls of Congress, or to a hurricane survivor's flooded home.

Your stories could break news, expose malfeasance, or convey a compelling narrative. No matter what, you will be doing some of the most consequential work in media. If journalists are responsible for writing the first rough draft of history, climate journalists are the only ones whose final story will be inscribed in the geologic record.

30
Conservation and Wildlife

Rachel Nuwer

Rachel Nuwer is a freelance journalist who often writes about conservation, ecology, and wildlife for the New York Times, National Geographic, Scientific American, *and other major media outlets. She is author of the award-winning book* Poached: Inside the Dark World of Wildlife Trafficking, *the reporting for which took her to 12 countries. She hails from Mississippi but now calls Brooklyn home, where she shares an apartment with a large orange cat and a harlequin bunny.*

People often use the term "rocket science" to describe something excessively difficult and complicated. In fact, they'd be better off using "conservation science" to make this point. Rocket science is dictated by a set of known rules about objects and motion. Conservation science, on the other hand, involves not only untangling extremely complex relationships between species and the environment, but also grappling with human behavior—one of the most unpredictable variables in the universe. As a source of mine once quipped, sending rockets to the moon and back has turned out to be way easier than trying to protect our planet from humans.

As a reporter covering conservation and wildlife, your job is to clearly capture and convey this complexity to readers—and to do so in just several hundred to a few thousand words! This is far from an easy task, but it's one that is of critical importance. Your stories have the potential to shift conversations, actions, and policy, resulting in changes that could literally save a species from extinction or a habitat from destruction. Media attention, for example, played a major role in getting countries to ban the international trade of ivory in 1989, and it also contributed to China's decision to close its domestic ivory market in 2018—actions that translated into a significant easing of poaching pressure on elephants in Africa.

Such high-level victories are, of course, exceptions rather than the norm in terms of what you can expect to come of your reporting. But they demonstrate the power and value of writing about conservation and wildlife. This beat is often crushingly depressing, seemingly mired in nothing but bad news. But

Rachel Nuwer, *Conservation and Wildlife* In: *A Tactical Guide to Science Journalism.* Edited by: Deborah Blum, Ashley Smart, and Tom Zeller Jr., Oxford University Press. © Oxford University Press 2022. DOI: 10.1093/oso/9780197551509.003.0031 you can stay motivated by focusing on the fact that you are giving a voice to otherwise voiceless species and habitats, the health of which is inextricably linked to our own.

Know Your History

Conservation biology is a relatively young field, having crystalized into a distinct discipline only in the last 50 years. But plenty has happened since then, and more still happened in the century-plus leading up to the field's emergence. Getting up to speed on the history of conservation will help contextualize your reporting and writing by lending insight into the various schools of thought that define current topics and practices. Hot-button debates include, for example, fortress conservation (in which militaristic park rangers protect land from all human incursion) versus sustainable use (in which local people are permitted to benefit and extract resources from the land in a way that does not harm it); and top-down management (in which the government or other officials call the shots) versus bottom-up management (in which the people living nearby make and enforce the rules).

It's also important to note that, like much of science, early conservation was mired in racism, the reverberations of which are still felt today. Michelle Nijhuis's recent book *Beloved Beasts: Fighting for Life in an Age of Extinction*, is a great crash course in some of the major players and developments that got us to where we are today.

Knowing your sources' history and viewpoints is also helpful. Besides just asking them what they think, you can glean insight into their opinions ahead of time by consulting their published scientific papers, searching for news stories they've been quoted in, reading blogs they may have authored, and scrolling through their Twitter feed.

In an ideal world, researchers adhere to what the data say, but you'll find that in conservation, especially, sources can be highly opinionated when it comes to the "right" way to protect the environment. Again, unlike in rocket science, in conservation there often aren't clear-cut answers about how to go about best protecting a species or habitat in an imperfect, human-dominated world. Consequentially, strong differences in opinion are commonplace. Trophy hunting, wildlife farming, and the legal trade of things like ivory and rhinoceros horn are just a few of the topics that regularly elicit strong debate. Figuring out which camp a particular source belongs to can help you determine whether they're right for your story and what sorts of biases or opinions they might hold.

Dig Deeper

More than a decade ago, when I first started investigating wildlife poaching as a student, I held the painfully oversimplified, black-and-white view that all poachers were bad people who deserved to be heavily fined or sent to jail. It took meeting a poacher firsthand in Vietnam and learning about his struggles to support his family for me to realize that my moral judgment was not only naïve, but harmful. While energy should clearly be invested into combating illegal hunting, focusing on locking up poachers—who are usually the sole breadwinner in an impoverished household—is not the solution.

My own reporting has led me to that conclusion, but it's also a view that is held by the broader conservation community. We should instead concentrate on eliminating poaching's main drivers, including poverty, corruption, demand for wildlife products, and higher level organized criminal activity. Unfortunately, in many places, policy and action on the ground have not caught up with the need for a more nuanced strategy.

The oversimplification trap occurs across the spectrum of wildlife and conservation stories, from gushing headlines about drones saving rhinos in Africa (they aren't) to claims that euthanizing man-eating tigers in India threatens the survival of the species (which it doesn't—tigers, if given space, prey, and protection from poaching, reproduce like cats). But armchair critics abound, and inaccurate, unhelpful messaging about conservation—for example, calls for blanket bans on trophy hunting in Africa—is frequently pushed by celebrities and others with no expertise in the subject.

To avoid contributing to shallow, flawed, or one-sided coverage, try to approach each story—especially those on a controversial topic—with an open mind. Put aside any preconceived notions you have and embrace the complexity that tends to accompany wildlife and conservation. Be aware, too, of your sources' motivations. A source from a nongovernmental organization or charity, for example, might have a specific message to push, one that could be informed by advocacy and fundraising rather than data and facts. Always try to speak to at least one outside source who does not have a direct stake in whatever it is you're reporting about, but who can speak with authority on the subject at hand.

Diversify Your Voices

Fisheries expert Francisco Blaha, reacting to the recent Netflix documentary *Seaspiracy*, about the environmental impact of fishing, recently complained on Twitter, "I'm over the setup where the 'bad guys' are predominately Asian, the 'victims' predominantly Black/brown and the 'good guys' talking about it and saving the ocean are predominantly white." This isn't a problem unique to fisheries-related reporting. The same "cliché stereotypes and racist overtones," as Blaha put it, bubble up across much of conservation and wildlife-related storytelling.

A straightforward way to avoid contributing to tired, problematic stereotypes is to go out of your way to include diverse voices from the place you are reporting about. If you're writing about pangolin-scale trade in Vietnam, interview Vietnamese conservationists—not just Western ones—about why this is a problem and what they're doing to stop it. If possible and applicable within the scope of your assignment, also try to conduct interviews with more than just scientists.

Academic experts' views are important, but given the human element of conservation, they represent just one piece of the puzzle. The types of relevant players you reach out to depend on the story, but the list could include government officials, activists, business representatives, community leaders, and consumers. Strive, too, to try to include sources of diverse racial, ethnic, and gender backgrounds. They may shed light from a different point of view on the topic you are writing about. Practically speaking, some newsrooms have also started to require female or minority voices in the stories they publish. Even if they don't, though, this is a good practice for any journalist to follow.

Don't forget, either, the people whose lives are affected by whatever it is you're writing about. Their anecdotes will add depth and humanity to your story and help you avoid generalizations. You may even wind up indirectly helping your sources. For example, in a *New York Times* piece about donkey theft in East Africa, I included the story of a Kenyan father whose donkeys had been stolen, costing him his job. A reader was so moved that she purchased new donkeys for him. This didn't solve the problem of donkey theft across the region, but it vastly improved my source's life. If nothing else, for me, that made the story worth telling.

Traveling to the place you are reporting about, especially if the location is somewhat remote, is the most effective way to interview a broad swath of sources. Given the reality of ever-shrinking newsroom budgets, this isn't always possible. But you can still reach diverse sources by email, phone, WhatsApp, Instagram, or Facebook. If language is an issue, ask your editors if they have a budget for translation and explain why it would be a boon to shell out funds. In some clear-cut, ethically unambiguous cases, you can also ask your English-speaking sources to help facilitate interviews and translations with other sources. When I wrote for *National Geographic* about grassroots fisheries reserves in Thailand, for example, the American scientist who led the research texted my questions in Thai to villagers he knew from his work and then translated their replies back to me.

Incidentally, the same advice about including diverse sources also applies the types of wildlife you decide to write about. There is more to life on the planet than elephants, rhinos, lions, and tigers. Some of the most imperiled plants and animals in existence are ones that most people have never heard of. Your story about an obscure turtle or overlooked songbird could be the catalyst that finally brings the crucial attention needed to better protect it.

Seek Out a Broad Audience

I sometimes feel like I'm preaching to the choir, that my stories are only being read by people who already care about wildlife and who already know all about conservation. By finding creative ways to package stories about these subjects, however, I find that I can reach new readers. I've written, for example, a culture story about traditional Japanese musicians trying to phase ivory out of their industry; a travel story about conservation victories in Chad; a tech story about new software being used to monitor national parks in Africa; a fashion story about high-end Italian, French, and U.S. designers' use of illegal wildlife products; a food story about a Zimbabwean chef who cooks vegan cuisine for an all-woman ranger team; a political story about changes to hunting legislation in Alaska; and a health story about the benefits of being in nature.

Unique, fresh angles for sharing news about wildlife and nature conservation are readily available if you keep an eye out for them. If you are a freelancer, this tactic will also help you land more assignments. There are only so many media outlets that accept straight-up wildlife and conservation stories.

It's Not All Doom and Gloom

When I started writing the proposal for my book, *Poached*, my agent specifically told me I had to make it hopeful. At the time, I thought this was wishful thinking. Illegal wildlife trade, after all, is a multibillion-dollar contraband industry that impacts thousands of species, but that receives relatively little attention compared to other forms of crime, like drugs and arms trafficking. How could there possibly be anything to be hopeful about?

Once I dove into reporting, however, I was pleasantly surprised to find there was plenty to be optimistic about, from the fact that people around the world (including politicians, business leaders, influencers, and ordinary citizens) are simply talking about wildlife trade way more today than they were 15 years ago, to the recent ban on consumption of many types of wildlife in China in reaction to the Covid-19 pandemic. None of these are complete solutions, but they are steps in the right direction.

Beyond the wildlife trade, there is even more to be excited about. Wonderful examples of progress that we can point to from around the world range from rivers being given legal rights in places as varied as Bangladesh, New Zealand, and Ecuador, to headway being made in the United States on a bill that would ban private big-cat ownership. While most of what I write about does veer toward doom and gloom, I love taking a break to cover the latest positive conservation story, whether it be an island ecosystem's speedy recovery following the removal of invasive rats or the installation of playfully inventive canopy bridges to help slow lorises, gibbons, and other tree-dwelling species safely cross roads.

The personal stories of the people who are dedicating their lives to saving animals and their habitats are also incredibly inspirational and heartening. One of the most surprising things I've learned while covering this beat is that conservationists are an unexpectedly optimistic bunch. As Tim Tear, the former head of the Wildlife Conservation Society's Africa program, once told me, conservation victories are possible because they have to be. "We don't have all the answers," he said, "but pockets of success demonstrate that we *can* figure out how to have successes."

While I tend to lean toward pessimism, I try to keep Tear's encouraging message in mind—that we can succeed, if only we care to try—and convey it to readers.

The Takeaway

Getting people to pay attention and care about wildlife and conservation often feels like a long, depressing slog, and as a result, this beat can be a frustrating corner of science journalism. But this doesn't mean that these challenging stories are worth skipping over in favor of a lighter, more popular read—quite the opposite.

As with journalists who report primarily about human subjects, those of us who write about wildlife and nature give a voice to the voiceless, hold power to account, and strive to right injustices. More than that, we bring attention to the wondrous diversity that defines life on Earth. And as numerous case studies show, there is some reason to be optimistic, and successful conservation is possible.

The first step to protecting a species or place, though, is to make people aware of its existence and plight and to make them care about it. That's where you, the conservation and wildlife reporter, come in.

31

Earth Sciences

Betsy Mason

Betsy Mason is a freelance science journalist based in the San Francisco Bay area. Her work has appeared in numerous publications, including National Geographic, Scientific American, Science, Nature, Science News, *and the* New York Times. *Previously she was senior editor in charge of online science coverage for* Wired *magazine and the science and national laboratories reporter for the* Contra Costa Times *(RIP), where she won the American Geophysical Union's David Perlman Award for her coverage of earthquake risk in the Bay Area. She was a Knight Science Journalism fellow at the Massachusetts Institute of Technology in 2015–2016 and is the secretary of the board of the Council for the Advancement of Science Writing. Betsy also writes about maps and cartography and is coauthor of the book* All Over the Map: A Cartographic Odyssey. *Before becoming a journalist, she earned a master's degree in geology from Stanford University.*

If you're browsing this guide hoping to find an underappreciated gem of a beat in science journalism, you've landed on the right chapter. I might be biased (being a former geologist and someone who genuinely loves rocks), but covering earth science has definite advantages.

Unlike some of the other beats in this guide (I'm looking at you, medicine), this one's not overflowing with journalists, so there's plenty of room. One reason for this relative scarcity could be that earth science encompasses many disciplines, and the need for a broad grounding in the sciences (including physics and chemistry, yipes!) can be daunting. But like a good job applicant, earth science would tell you this perceived weakness is actually a strength: The beat is an excellent opportunity to gain that knowledge base while on the job.

Earth science has a little of everything, from geology and geodesy to oceanography and meteorology. There are the marquee topics, like destructive earthquakes, erupting volcanoes, and the latest clues to whether *Tyrannosaurus rex*'s giant teeth and tiny arms are hallmarks of a ferocious

Betsy Mason, *Earth Sciences* In: *A Tactical Guide to Science Journalism*. Edited by: Deborah Blum, Ashley Smart, and Tom Zeller Jr., Oxford University Press. © Oxford University Press 2022. DOI: 10.1093/oso/9780197551509.003.0032 hunter or a benign scavenger. But there are also scores of under-the-radar areas, littered with potential scoops.

"There are so many opportunities here," says Eric Hand, *Science*'s European news editor and an experienced earth science reporter. "You're never going to be out of a job." (OK, we can't guarantee that.)

And let's not forget the fieldwork. Many earth scientists spend a good chunk of their time outside, often in amazing places like the Italian Alps, Baja California, or Russia's Kamchatka Peninsula. As an earth science reporter, you could find yourself tagging along with researchers to Antarctica or on a deep-sea research vessel. Even if you don't get into the field more than once in a Milankovitch cycle, you still have the chance to report on every corner of the globe.[1]

Another bonus is that earth scientists tend to be a laid-back bunch. "They wear ball caps and drink beer in the field," says Dick Kerr, who recently retired from *Science* after 37 years on the beat. "I found them very approachable." Convinced? Great. Here are some things to keep in mind.

The Hazards of Reporting on Hazards

Most things in earth science happen on *very* long timescales: We're talking thousands, millions, even billions of years: plates move, mountains erode, glaciers creep, species evolve. But these gradual processes are punctuated with sudden, spectacular, and sometimes deadly events like earthquakes, tsunamis, eruptions, and landslides.

These natural hazards make for exciting stories, but they also require extra care and consideration. "Avoid the whole disaster porn thing," says Alex Witze, a veteran freelance journalist who covers earth science and is quick to joke that she herself is guilty of writing an entire disaster porn book about Icelandic volcanoes. (*Island on Fire*, which she wrote with Jeff Kanipe, is actually a great blend of science, history, and storytelling.)

"Maybe it's cool to have a nice anecdote on the ground, but what would it be like to be the family member of that person who's in your lede who maybe died?" Witze says. "Just think about the effects of your writing, especially while you're reporting in the heat of the moment."

---

[1] Milankovitch cycles describe the effect on the climate of small changes in various Earth movements, like the shape of its orbit and its tilt relative to the orbital path. They are on the order of tens of thousands or hundreds of thousands of years.

Most reporting on natural hazards isn't in the immediate aftermath of a catastrophe, however, and it can be tempting to play up the risk of the next major event. During my time as the science reporter for the *Contra Costa Times* in the fault-ridden San Francisco Bay Area, I felt that part of my job was to make sure my readers knew that every one of them was living within striking distance of a potentially deadly fault. But doing that effectively, and responsibly, requires careful consideration.

The Bay Area's most infamous fault may be the San Andreas, but many geologists and seismologists think the Hayward fault, which runs alongside the densely populated cities on the eastern side of the bay, including Berkeley and Oakland (and right through downtown Hayward), is more likely to cause the next major disaster. Geologists estimate that the average time between major quakes on the Hayward fault is around 140 or 150 years. The last one was in 1868. So we're overdue, right?

That's one way to look at it. But the time between events has varied from less than 100 years to well over 200 years. It could be another century before the next rupture. Or it could happen tomorrow. In reality, the chance that a major quake will occur on any given day is very, very small. But scientists place the odds that one of the Bay Area's faults will unleash a destructive quake over the course of a typical 30-year mortgage at better than even. Putting this sort of risk into perspective for readers is not an easy task.

The best course of action is to talk to as many scientists studying different aspects of this risk as possible. It's part of the U.S. Geological Survey's mission to help people prepare for natural disasters, so that's a good place to start. You can interview seismologists, geologists, seismic engineers, public policy experts, and behavioral scientists who research how people respond to risk, to name a few.

Understanding the reason an area is prone to disaster can also help readers wrap their minds around the risk. The Bay Area straddles the seam between two enormous tectonic plates that slide horizontally past each other at a rate of around 5 centimeters per year. Friction keeps the plates locked together until the stress of that motion builds up enough to cause the sides to violently snap past each other (as much as 20 feet or more in the 1906 earthquake) like an overstretched rubber band that finally breaks. It's the science journalist's job to explain those underlying forces.

Rock the Beat

Stretching from Earth's core to the crust of Mars and beyond, the expansive nature of the earth sciences yields countless types of stories and plenty of undercovered areas. Much of this territory has the benefit of being, well, down to earth.

"Things in earth sciences are at least to some degree familiar to any reader," Kerr says. Stories about glaciers or the seafloor tend to be more tangible than stories about dark matter or mouse genomes. This lends earth science stories some natural appeal.

Beyond covering natural hazards, you'll find stories with a biological bent, like mass extinctions in the fossil record, or the health impacts of a smoke plume following a wildfire. Many earth science stories have a business angle, including earthquakes caused by fluid injection at oil and gas wells in Oklahoma or the distribution of rare earth elements critical for products like smartphones. The broad range of the beat means there are lots of possible outlets for stories.

News pegs for earth science stories can be found in all the regular places. Happily, earth scientists seem more willing to present newsworthy findings at scientific conferences, like the annual meeting of the American Geophysical Union, rather than keeping them under wraps until they're published in a journal.

Like any beat, news can be pegged to journal publications or posts on the relatively new EarthArXiv preprint server (see Chapter 3 on journals, peer review, and preprints for more on this). Events like a planetary flyby, the launch of a new remote sensing satellite, or the anniversary of a major earthquake are also good pegs. (I took full advantage with a deluge of stories on the 100th anniversary of the 1906 quake.) Tagging along with a scientist in the field can be news peg enough for some editors, and these stories are also great opportunities to write about scientists in action.

Another great selling point for many earth science stories is that their subjects can be quite photogenic. "Often getting really great visuals is easier and helps make for a more compelling story," says Hand, who recently wrote a story for *Science* about atmospheric gravitational waves that included a ridiculously scenic photo of the lenticular "UFO" clouds associated with those waves hovering over the eye-catching Torres del Paine mountains in Patagonia at sunset. "I made the art team happier than they'd been all month."

Earth science shares fuzzy borders with several other beats, and while some toes may be stepped on now and then, this overlap means more opportunities. On the (disputed) border with the space beat, you'll find great stories about planets. Much of what we know about other rocky planets is based on our knowledge of Earth. Like movie location scouts, some planetary geologists hunt for places on Earth they can study as analogues for extraterrestrial terrain, like ice-sealed Antarctic lakes as a proxy for the ocean beneath the icy crust of the Saturnian moon Enceladus. This makes planets and moons natural subjects for earth science reporters.

The same goes for other border beats, like environment, where you'll find stories about the potential negative impacts of fracking sludge pits and mountaintop removal mining or stories about shoreline erosion and land subsidence. In the climate change realm, the earth science journalist might tackle stories about the clues to past climates found in ice and sediment cores or the potential to use the carbonate rock cycle to pull carbon back out of the atmosphere.

Back squarely in the middle of the earth science beat is an amorphous category of stories Witze affectionately calls "weird earth stuff." This includes a popular story she wrote for *Nature* about how Earth's magnetic pole is moving faster than expected. "I think that the weird, wacky planet has an inherent kind of attraction for most editors," she says.

Another good example is a story by Maya Wei-Haas, who covers earth science at *National Geographic*, about an odd, low-frequency seismic wave that was ringing around the globe with no apparent source. She picked up on the story on Twitter, where scientists were discussing the mystery. And she wrote a follow-up story when scientists traced the waves to a baby volcano developing near the tiny islands of Mayotte between Africa and Madagascar. "It was a pretty neat story as a whole," she says.

On the fringes of the beat there's lots of relatively untrodden terrain (and plenty of scientists who'd be thrilled if a journalist showed interest in their work). Scant attention is paid to the deep interior of the planet or the edge of the atmosphere, where there are stories like one I wrote for *Science* about unexpected disturbances to a stratospheric wind, known as the quasi-biennial oscillation, that could impact global weather patterns and make seasonal forecasts less accurate.

One reason journalists may steer clear of the beat's fringes is that you're likely to encounter some daunting science there, like fluid dynamics and atmospheric chemistry. Hand agrees that these subjects can be challenging, but if you're ready to "dive into the muck and figure it out," there are good stories to be had. "I don't think it's any different from covering anything else," he says. "If you're willing to ask questions and ask questions until at least you think you have a reasonable understanding, you'll be fine."

Watch for Falling Rocks

Like all science journalists, those covering earth science need to beware of words that have a generally understood lay meaning but have a different scientific meaning. For example, people tend to use prediction and forecast interchangeably, but for earth scientists these words have specific meanings. An earthquake forecast gives a *probability* of an event happening in a given time window in a specific area within a specific magnitude range. Nobody can predict an earthquake, which involves a statement that a particular event *will* happen at a certain time and place. Likewise, it's important for people to understand that a "100-year flood" is not a flood that happens once every 100 years. It actually means there's a 1 percent chance of a flood of a certain size happening in any given year.

Earth science reporters also need to be cognizant of the effect sensationalized stories about things like near-earth asteroids, magnetic pole reversals, earthquake swarms, or so-called supervolcanoes can have on readers. "Some of these scary headlines that come out sound like it's a planet-ending affair that's happening," says Wei-Haas. "I think it's really important to understand that truly does cause people to freak out. It might get you a lot of clicks on a story, but it does a lot of damage at the same time."

Taking the time to grasp the context of the research and carefully explain it to readers is critical, Wei-Haas says. For all things seismic, she suggests journalists start with the explainers on the website of the Incorporated Research Institutions for Seismology. Likewise, the Smithsonian Institution's Global Volcanism Program has background and updates on any volcano in the world. And for boning up on earth science in general, the U.S. Geological Survey has a wealth of information. A lot of the goodies on these websites don't show up in a Google search, she cautions, so you'll need to search on the websites themselves.

Context is especially important for geology stories because, unlike biology, physics, and chemistry, most readers likely never studied it in school. Familiarity with mountains does not necessarily come with even the slightest understanding of how those peaks got there. "Earth science reporters have to be particularly careful not to presume too much about what the reader knows about the planet," Kerr says.

A final important point to keep in mind is that earth science is among the least diverse scientific disciplines, with only 10 percent of degrees going to scientists of color and a seriously leaky pipeline for women ascending the ranks. It's worth the extra effort to seek out diverse sources for your stories. Wei-Haas suggests taking advantage of the local partnerships Western earth scientists sometimes forge when doing fieldwork in places like the Himalayas or Mongolia. Another way to do it is to skip the interview with the senior scientist in charge of the research group, who's more often than not a man, and call the woman who was actually in the field hitting rocks with a hammer.

The Takeaway

The earth science beat is filled with underreported topics and offers many avenues for intrepid science journalists who don't mind getting their hands (or boots) dirty tackling a wide range of scientific disciplines. Reporting on natural hazards is a great opportunity to do meaningful service journalism that might even save lives, so it's important to provide the proper context for readers and avoid sensationalizing these stories.

32

Mathematics

Jennifer Ouellette

Jennifer Ouellette is the senior writer covering science and culture for Ars Technica *and the author of four popular science books:* Me, Myself, and Why: Searching for the Science of Self *(2014);* The Calculus Diaries: How Math Can Help You Lose Weight, Win in Vegas, and Survive a Zombie Apocalypse *(2010);* The Physics of the Buffyverse *(2007); and* Black Bodies and Quantum Cats: Tales From the Annals of Physics *(2006). She also edited* The Best Online Science Writing 2012 *and is the former science editor of* Gizmodo. *Her freelance work has appeared in* The Washington Post, The Wall Street Journal, the Los Angeles Times, The New York Times Book Review, Discover, Slate, Salon, Smithsonian, Mental Floss, Pacific Standard, Alta, Nature, Physics Today, Physics World, Quanta, *and* New Scientist, *among other outlets.*

One day many years ago, my husband and I were browsing in a Santa Barbara, California, bookstore and wandered over to the science section. I had just started writing a book about calculus; this was from my perspective as a former English major and recovering mathphobe. My spouse, a physicist, handed me a book he thought might prove useful to my research. I opened it to a random page and was confronted with the sight of multiple equations scattered throughout the prose. I snapped it shut and handed it back with a shudder: "Oh, no, not *that* one!"

Such is the challenge facing any writer who dares to take on the task of communicating a math-intensive story to general readers. Most people recall their high school math classes as being boring, frustrating, confusing, or downright traumatizing. We memorized the rules and dutifully crunched the numbers by rote, all while fighting off an unrelenting sense of dread lest a tiny error along the way produce the wrong answer. For so many people, math class was the one that most damaged their self-confidence, lowered their grade point average, and made them feel stupid. No wonder, as adults, they opt to avoid math whenever they can.

Jennifer Ouellette, *Mathematics* In: *A Tactical Guide to Science Journalism*. Edited by: Deborah Blum, Ashley Smart, and Tom Zeller Jr., Oxford University Press. © Oxford University Press 2022. DOI: 10.1093/oso/9780197551509.003.0033

Even seasoned science writers have been known to balk at tackling a story that involves a lot of math. I was literally writing a book about math, and had been writing about physics regularly for 15 years, when I slammed shut that book in Santa Barbara.

In short, math's reputation precedes it. The good news is that this is learned behavior, and that means it can be unlearned. Deep down, most people are curious about the world around them, including math. They want to understand it; they just think they can't based on past negative experiences. The trick is to find clever ways to sneak past that knee-jerk defensiveness.

The First Rule of Math Club

When the physicist Stephen Hawking published his bestselling 1988 book *A Brief History of Time*, he noted in the acknowledgments that his editor had persuaded him to remove all but one equation: $E = mc^2$. The rationale was that, for every equation, the readership would be cut in half. Hawking's book ended up selling over 25 million copies, so it proved to be wise counsel. That maxim has been accepted wisdom ever since—call it the "first rule of math club." But it poses a daunting challenge for anyone wishing to write about math.

All disciplines have their share of dense jargon, and a large part of being a good science writer is figuring out how to translate highly technical material into plain language. Mathematics is a rich symbolic language all its own—*and* it also has obtuse technical jargon on top of that. All those symbols might as well be ancient Sanskrit or Egyptian hieroglyphics to the average reader: How do you even pronounce å, anyway? Small wonder that so many people run screaming in the opposite direction at the mere sight of an equation.

Ben Orlin, a math teacher best known for his *Math With Bad Drawings* blog, generally follows the conventional wisdom and avoids mathematical symbolism in his popular books. "Mathematical equations are the tersest, densest language we have, hard to make sense of even if you put in the work," he says. "If you've got crucial information to convey to a reader, an equation is a great place to hide it so no one finds it."

There are exceptions to the rule. Stanford physicist Leonard Susskind's 2014 book, *The Theoretical Minimum: What You Need to Know to Start Doing Physics*, was rife with equations, and yet it briefly became a *New York Times* bestseller, prompting two sequels. But Susskind's book was based on his series of equation-intensive YouTube courses on modern physics that had garnered a devoted following among hard-core physics enthusiasts—most of whom had sufficient background and passion for the topic not to be put off by equations.

There's certainly an audience for a higher level of science writing, pitched somewhere between technical papers, monographs, and textbooks on the one hand and articles and books written for general readers with little science or technical background on the other. And equations do have a certain aesthetic appeal, as photographer Jessica Wynne discovered when she began taking pictures of the chalkboards of mathematicians around the world in 2018. She likened the aesthetics to abstract paintings and found that each mathematician had their own style of calculation, making their chalkboards abstract portraits of their personalities.

That innate aesthetic appeal is why I sneakily included relevant equations in the illustrations of my calculus book as a way to gently acclimate math-phobic readers to the sight of them. But if you really want to reach those who aren't already part of the mathematical fold, eschewing equations is still a good rule of thumb.

That said, "You feel very handicapped if you can't show equations, because that is our currency," says Steven Strogatz, a Cornell University mathematician who has written several popular books about math, as well as a popular series of columns for the *New York Times* in 2010. "We're used to expressing ourselves with proofs or equations."

As a guideline, Strogatz finds it helpful to envision two axes in a simple graph. The horizontal axis is concrete/abstract things, and the vertical axis is whether something is familiar/unfamiliar to a general reader. The trick is to strike a balance between the two.

For instance, for one of his *New York Times* columns, Strogatz couched a discussion on the tricky concept of conditional probability—the likelihood that a particular event happens given the occurrence of another event—within the familiar context of the probability that a woman with a positive result on her mammogram actually has breast cancer. Math students are typically taught to apply a complicated formula to calculate the odds, but Strogatz suggested a simpler method: rephrasing the various percentages and probabilities—for example, a 0.8 percent probability that one woman in a group of 1,000 has breast cancer—in terms of natural frequencies (8 out of every 1,000 women have breast cancer). It's a shorter, more intuitive approach that might lack the rigor of the traditional formula, but clarifies the problem for the average person to help them arrive at the correct answer (9 percent).

For another column, Strogatz walked readers through the mathematical proof for the area of a circle—an abstract concept couched in a familiar shape. He stumbled, however, in a column about curved spaces and how they impact the familiar maxim that the shortest path between two points is a straight line, including a side discussion of more exotic shapes like a two-holed torus. "I over-taxed my reader," he admits. "You can do unfamiliar and concrete, and you can do familiar and abstract. But you cannot, I learned, do both unfamiliar and abstract."

The concrete and familiar are my mathematical playground as a science writer. There is a rich array of examples to choose from because math is everywhere. It lurks in your knitting patterns, in knots and smoke rings, in the exquisite architecture of Antoni Gaudi, and in the trajectory of water spurting from a sprinkler system. Math is there when a surfer catches a wave in Hawaii, when a gambler calculates his odds of winning at craps, or when a child swings back and forth in a playground. Math can show you how to brew the best espresso or find the best parking space in a crowded lot. Every time we comparison shop, we're effectively solving an optimization problem. The TV series *Numb3rs* put it best: "We all use math every day."

Every Equation Tells a Story

Still, everyone who writes about math will inevitably one day find themselves faced with trying to write clearly and coherently about an esoteric topic for which nothing short of an equation or two will suffice to do it justice. "For many ideas, it's the only language we have," Orlin admits. "And even for ideas that *can* be conveyed by other means, there's a force and clarity and concision to mathematical symbols, not to mention an iconic power."

Take the well-known dictum in physics that all objects fall at the same rate, regardless of mass. This contradicts our personal lived experience since if we drop a feather and a coin at the same time, we will see the coin hit the ground well before the feather. This is not because the physics is wrong, but because there is a confounding factor: friction from air resistance, which slows the feather's descent because it has more surface area than the coin. Conduct the same experiment in a vacuum, and both objects really do fall at the same rate.

Or you can simply do the math. Walk yourself through a simple algebraic equation, and you'll see how the little m—representing the mass of the object—on each side of the equation cancels the other out, making an object's mass irrelevant to the rate of acceleration.

So how can you determine whether an equation is required? "Every equation should be a statement telling a key part of the story," Orlin says. When deciding whether to incorporate an equation or not, he follows a simple rule of thumb: "Am I willing to spend a paragraph or two explaining, term by term, what each symbol in this equation means? If not, then maybe it isn't necessary."

I like to say that every equation tells a story; it just does so with numerical symbols instead of words. When Abu Jafar al-Kwarizmi invented modern algebra in the first century, he expressed his unknowns in words rather than variables and his equations in sentences. In essence, that is what a mathematical equation is: a sentence reduced to a symbolic shorthand so the quantities can be more easily manipulated.

That's why, when breaking down equations for general readers, Orlin suggests using words instead of symbols, such as typing out the word "wavelength" rather than using the traditional Greek letter lambda (λ). In his mind, equations play two fundamental roles: one, to convey information, just like any other sentence; and two, to allow us to manipulate the symbols to perform calculations. "If you are writing for a general audience, you can usually avoid that second role, and focus on how the equation enhances your narrative," he says.

Tell us a good story and we'll follow you anywhere, even into the minefield of scary equations. The narrative potential of even the most esoteric mathematical concepts is boundless because math is very much a human endeavor. Within every equation lurks a rich history, providing ample opportunities for anecdotes, colorful characters, historical details, and various subplots to ornament the fundamental concept you're trying to communicate to readers and keep them engaged.

As an example, let's break down Einstein's $E = mc^2$. E represents energy, m represents mass, and c^2 is the speed of light squared. Einstein's revolutionary insight was that energy and mass are equivalent and interchangeable. Mass can turn into energy and vice versa, and the speed of light squared is the exchange rate between them. Break apart an atom and the fragments will have slightly less mass than the original atom because a certain amount of mass was changed into energy.

The speed of light squared is a very large number, so a small amount of mass can produce a tremendous amount of energy, although it is released gradually. Still, it wasn't long before physicists built on Einstein's insight to figure out how to trigger a chain reaction to release that energy all at once, giving us the Manhattan Project and the dawn of the nuclear age and the multitude of human stories contained therein. As novelist Jeanette Winterson observed in her novel *Gut Symmetries*, "Inside the horror of Nagasaki and Hiroshima lies the beauty of Einstein's $E = mc^2$."

Doing the Math

What about showing actual derivations—the painstaking, step-by-step process of working through an equation to find a solution? Strogatz advises against it. "I think you really have to avoid long derivations, or sequences of equations that make logical arguments, because [a general audience] won't follow it," he says. It's good advice in general. But I would argue that this, too, can work in special cases if handled carefully.

I avoided most equations and all derivations in the main text of my calculus book, but I also included two appendices for that rare reader sufficiently intrigued to dare to dip a toe into the actual math. For example, I adapted a textbook problem I'd worked on during my research phase into a fun, familiar context: zombie movies, or what I called the "calculus of the living dead."

I walked the reader through each step as I derived a simple equation ($y = Ce^{kx}$) to answer a practical question: Given the known rate of infection k, how soon should you evacuate once the first zombie shows up in your town? C stands for the initial population of zombies (it's a constant number that does not change), while y denotes the total number of zombies after a certain number of days, denoted by x.

If we know that $C = 19$, $y = 193$, and $x = 10$, we can calculate the rate of infection: $k = 0.231825$. Now we can use this equation to predict the number of zombies after any amount of days simply by changing the value of x. After 30 days, for instance, there would be nearly 20,000 of the undead—a classic case of exponential growth. Real epidemics have many more confounding factors, but the basic principle still stands.

Rhett Allain, a physicist at Southeastern Louisiana University, is a master of successfully deploying equations and, on occasion, detailed derivations in his articles for *Wired*. Take his analysis of a scene at the end of *The Empire Strikes Back*, when the *Millennium Falcon* makes a dramatic jump into hyperspace. Based on data gleaned by tracking the scene, and aided by the occasional diagram, he demonstrated how to calculate the various physical forces acting on the passengers during the jump. If you find you really need to include a calculation, Allain's work provides a useful model of how to do it exceptionally well.

Ultimately, it is worth the effort to occasionally navigate the minefield of incorporating equations. "We should help readers grasp mathematical symbols and equations precisely *because* they often act as a barrier," Orlin says. "The ability to read math is a kind of literacy, and telling readers, 'Eh, you don't have to worry about this,' consigns them to illiteracy. I don't want a world divided between a mathematical priesthood and nonmathematical masses, the former wielding mystical symbols and the latter cowering in fear."

Neither do I. Knowledge is power. Knowledge grants access to new intellectual territory and expands our range of choices in life. If we never, *ever* challenge our readers to face down their fear and grapple with equations, we are contributing to an ever-growing chasm between those who understand math and those who don't. In effect, we help disenfranchise them.

The Takeaway

Overall, the same general rules apply to writing about math as to any other scientific discipline. Avoid technical jargon and only use equations in rare cases. Make connections between the abstract ideas and real-world contexts familiar to readers. Visualizations and telling lots of stories can make even the most difficult topics palatable. Keep your sentences short and simple when you're breaking down the technical details and save your elegantly ornate phrases for the surrounding narrative framework.

The most critical thing is not to let yourself be intimidated by the subject matter. You don't need to avoid mathematical topics just because you weren't "good at math" and are afraid you won't be able to make sense of it.

We need to show general readers that those abstract symbols hold real meaning. The world is filled with hidden connections, recurring patterns, and counterintuitive truths that can only be seen through math-colored glasses. As Galileo once observed, "Nature's great book is written in mathematical symbols." It's up to us, as science writers, to make a compelling case to that effect.

33

Science Policy

Dan Vergano

Dan Vergano is an award-winning science reporter in the Washington, D.C., bureau of BuzzFeed News, where he has worked since 2015. He was formerly a senior writer and editor at National Geographic *and the senior science writer at* USA Today. *He was a Nieman Fellow at Harvard in 2007 and is on the board of the Council for the Advancement of Science Writing.*

"Every time you scientists make a major invention, we politicians have to invent a new institution to cope with it." (John F. Kennedy)

On February 3, 2020, Dr. Nancy Messonnier, a scientist with the Centers for Disease Control and Prevention (CDC) leading the response to the budding Covid-19 outbreak, announced some good news. Her agency had developed a test for the novel coronavirus and requested Food and Drug Administration (FDA) authorization for its use by public health labs.

"This will greatly enhance our national capacity to test for this virus," said Messonnier, before moving on to discuss quarantined travelers at an airbase. No journalist asked a question about the CDC's test during the briefing.

But embodied in that announcement was a disastrous science policy—essentially a decision to view the novel coronavirus outbreak as a likely repeat of the 2003 severe acute respiratory syndrome (SARS) virus that called only for rare testing by public health labs, meaning that other tests would not be needed. That policy would effectively blind the United States to the arrival of SARS-CoV-2 over the next 2 months.

This fundamental misjudgment was worsened by the contamination of the CDC's test, rendering the only approved tool for finding the virus useless. Only when the CDC allowed states to do their own tests was the early Covid-19 invasion revealed.

There's nothing remarkable about Messonnier's test announcement going unquestioned. Reporters routinely ignore most science policy news because they tend to see it as the grinding of governmental or industrial activity or

Dan Vergano, *Science Policy* In: *A Tactical Guide to Science Journalism*. Edited by: Deborah Blum, Ashley Smart, and Tom Zeller Jr., Oxford University Press. © Oxford University Press 2022. DOI: 10.1093/oso/9780197551509.003.0034

boring announcements of blue-ribbon panels and investments into promising research. But as the CDC's testing policy catastrophe shows, the policies behind these bland proclamations can have serious, even calamitous,
consequences.

The biggest science stories of this century (at least so far)—the Covid-19
pandemic and climate change—are riddled with science policy failures, to
use the term of art in policy analysis (yes, this exists) for screwups that hurt
people. Similar botches, and stories, abound in everything from building
space rockets to the opioid epidemic and nuclear strategy. Behind nearly all
of these failures are policy decisions involving science both done and left un-
done, determinations that shape the lives of readers and listeners in profound
ways that journalists need to bring to light.

What Is Science Policy?

Broadly speaking, covering science policy means reporting news about any
decision, official or unofficial, on technical matters (distinct from bread-and-
butter science news of discoveries) that affect the public. In a modern society,
where everything from landing a date to landing on Mars is a matter of scientific disagreement, such decisions are everywhere and can make critical
differences in the lives of your audience.

On the simplest level, science policy reporting is scorekeeping: how much
money goes to which federal agency for research, or announcements of
investments in labs by businesses, or the naming of luminaries to panels such
as the President's Bioethics Commission or the FDA's Vaccines and Related
Biological Products Advisory Committee. These research policy stories are
routinely covered in the news sections of outlets like *Science* and *Nature* and
by lobbying groups keeping track of grants.

For a science policy reporter serving a popular audience, these are the
walks and bunts of the trade, simple to report and newsworthy only when the
funding, project, or luminary is—despite everyone's best efforts—interesting.
The Trump administration excelled at making all three newsworthy, from
early threats to cut the budget of the $32 billion National Institutes of Health
by $5.8 billion (its budget is now $40 billion) to erecting a fetal research ethics
board and the naming of various fossil fuel industry advocates to environmental science positions (judicial overturning of their misfires will employ
environmental reporters for years).

These stories usually follow an annual cycle, beginning with each
administration's largely ceremonial budget proposals in the spring,

Congressional budget negotiations over the summer, and the votes on spending bills (maybe) in the fall, timed around the end of the federal fiscal year (FY) on September 30. Typically, the amount of funding, or the misstatements of controversial figures, are public knowledge; the opponents, supporters, and critics are obvious; and the issues to underline are clear. These bread-and-butter stories are good ones to dangle in front of an editor to demonstrate one is being diligent and best disposed of in a straightforward news piece, if done at all.

But they are also worth pitching, and doing a little reporting on, for reasons beyond entertaining your editor with your workaholism. "Personnel is policy" in politics, as the saying goes, and the naming of climate science opponents or advocates to the Environmental Protection Agency or National Oceanic and Atmospheric Administration not only sends a message to voters but also changes lives. The opposing views of President Trump's advisors on the need for the administration to embrace masking, for example, clearly made the U.S. response to the pandemic worse.

We are also talking real money in these smaller stories. The "science budget" in the federal budget for research and development (R&D) is more than $165 billion a year (Table 33.1)—as indicated in the table below—and U.S. business research funding is more than $400 billion.

Table 33.1 R&D in FY 2021 Appropriations by Type (Budget Authority in Millions of Dollars)

	FY 2019	FY 2020	FY 2021	FY 2021	FY20 Change		Request Change	
	Actual	Estimate	Request	Final[a]	Amount	Percent	Amount	Percent
Basic research	39,352	43,351	40,573	45,515	2,164	5.0%	4,942	12.2%
Applied research	45,692	46,911	40,839	2,451	1,334	2.8%	7,405	18.1%
Development	60,574	67,788		66,036	−851	−1.3%	902	1.4%
R&D facilities	4,359	6,005	3,818	4,766	−1,239	−20.6%	947	24.8%
Total R&D	149,977	164,056	151,267	165,463	1,408	0.9%	14,197	9.4%
Defense R&D[b]	72,004	80,506	76,478	80,773	268	0.3%	4,296	5.6%
Nondefense R&D	77,973	83,550	74,789	84,690	1,140	1.4%	9,901	112%

[a]AAAS's estimates based on OMB (Office of Management and Budget) and appropriations data.

[b]Includes Defense Department and other military agencies. The above figures do not reflect emergency Covid-19 R&D or the amended FY 2021 budgets for public health departments, such as the U.S. Centers for Disease Control. *Note.* The projected inflation rate between FY 2020 and FY 2021 is 2.0 percent.

All figures rounded to the nearest million. Changes calculated from unrounded figures.

Image credit: AAAS.

Most important, these stories signal to sources that you are following their interests, making it worth their time to seek you out and offer you more news. The basic theory behind beat reporting is that by covering these incremental policy stories, you are gaining the insights that will lead to better sources and deeper stories. There are countless examples: The George W. Bush administration's 2001 embryonic stem cell policy was blown up by science reporters asking about the number of cell lines available for research. A provisional CDC update showed that U.S. drug overdose mortality had risen in 2019 and upended Trump administration claims to have turned the corner on heroin and fentanyl deaths. Something tipped somebody to these stories.

Inside and Outside Stories

There's no single recipe for a great science policy story, but they all emphasize relevance, insight, and originality. The first quality is key: While scientific discoveries stories can rest on sheer wonder for their oomph, science policy stories typically have to work harder to connect to readers: "A top Trump appointee repeatedly urged top health officials to adopt a 'herd immunity' approach to Covid-19 and allow millions of Americans to be infected by the virus," begins a 2020 investigative piece by Dan Diamond in Politico, for example, that told readers about a dispute over herd immunity. Importantly, Diamond hit readers right away with why they should care.

Just like a science reporter explaining how a new discovery was made, the science policy writer must also illuminate how policies succeed or fail. It's not enough to say a decision was made; you have to explain why. Here's the way a 2012 story written by Brendan Maher in *Nature* explained how an obscure science panel ended up approving bioengineered viruses: "The most pressing question is why the research wasn't flagged up earlier for scrutiny. The answer: the policy simply wasn't in place." It takes a lot of reporting to arrive at this kind of simple insight.

Originality is a little easier because in-depth science policy pieces are so rarely attempted. Typically, flocks of news stories are written about a science-related controversy, without any larger explanation. A beat reporter should always be asking themselves, and their editor, if a story has reached the point of being ripe for a deeper dive.

A simple way to categorize science policy stories with all these qualities is to ask whether they come from "inside" science, such as researchers self-imposing a moratorium on recombinant DNA experiments in the 1970s, or "outside," such as President George W. Bush limiting stem cell research funding in the 2000s. Generally speaking, the outside stories will attract more readers—and awards.

Look Beyond Science Agencies

Too often, science reporters limit themselves to science agencies and presidential advisors. It's worth recalling that Operation Warp Speed, the pandemic's multibillion-dollar public–private partnership to create vaccines and treatments, involved military and industry figures, and their missteps made news, revealing mistakes in both the supply and distribution of vaccines.

Frankly, reporting policy news from either the military or industry is harder than from science agencies, but it may be more important: Decisions to pursue a new kind of hypersonic nuclear missile or drone-frying microwave weapons might lead to World War III. Internet platform algorithms, and the research they fund to pursue their fine-tuning, seem to likely play a part in how political advertising and civil strife proceed for decades to come. Google collected more than $83 million in advertising from each of the Biden and Trump campaigns in the last U.S. presidential election, for example, and Facebook spread propaganda in Myanmar's 2018 genocide campaign against the Rohingya.

Science policy reporting involving industry and military sources is especially difficult because they are able to hide information much more thoroughly than science agencies. The military's classification system is an excellent shield for keeping mistakes from Congress and the public (although some military information is still subject to open records laws). Industry hides behind laws protecting proprietary information—and for public companies, fear of angering the Security and Exchange Commission or litigious shareholders.

Follow the Science

Digging behind all these fronts takes more work and a different mindset. If in political journalism, the secret is to follow the money; in science policy reporting, the secret is to follow the science. The result is more scientific policy reporting rather than simply reporting science policy as described previously. That is to say, you are finding out if what your readers are being told measures up, using science as the ruler. If it doesn't, then ask why.

A perfect example came in 2010, when the *Deepwater Horizon* oil rig blowout killed 11 workers and started leaking what BP (British Petroleum) and the U.S. government said was 5,000 barrels of oil a day into the Gulf of Mexico, justifying a moderate cleanup policy. But on May 12, BP released a video showing oil flowing from the damaged well bore. Steven Wereley, an engineering professor at Purdue University, quickly analyzed the video, and NPR's Richard Harris reported the results, which indicated that the leak could be more than 56,000 barrels of oil a day. A week later, Wereley was testifying in front of Congress, and the failure to contain the spill became a national scandal that embarrassed the Obama administration for the rest of the summer.

While the justifications for decisions are endless, their scientific underpinnings often aren't—satellites orbit, toxins sicken, and projectiles penetrate—and the experts in these disciplines are the same scientists we rely on for stories of scientific discoveries. Talk to them about a new policy promising utopia the way you would check out a new paper in *Nature* concluding a fossil belonged to a fire-breathing dinosaur. In other words, be as skeptical of science policies as you would information in any other field.

Zombies of Science Policy

Just as every scientific discipline has a few shopworn ideas that can't seem to be killed—say, the insistence that Nobel Prize winners have worthwhile opinions about everything—science policy has a few hoary notions shambling around.

The first is an assumption that people no longer trust science because of the latest scandal of the moment, from the replication crisis to dodgy preprints released during the coronavirus pandemic. There's not much evidence for this. Americans trust science, by and large, and these sorts of disputes really don't make much of a dent in their faith in the people who brought them vaccines, protecting them from infections ranging from polio to Covid-19. A Pew Research poll showed that from 2016 to 2019, public trust that scientists were acting to benefit society grew from 76 percent to 86 percent, in the range observers have seen in such polls going back decades. Be very careful when scientists warn you off some sordid squabble because it will tarnish science's fragile halo. That's not your problem, and it won't happen anyway.

Another revenant in science policy reporting is the drumbeat for herding young people into STEM (science, technology, engineering, and mathematics) jobs, a claim conspicuously popular with universities, labs, and employers. In 2012, the President's Council of Advisors on Science and Technology said the United States would need to increase its yearly production of undergraduate STEM degrees by 34 percent to match the demand forecast for STEM l.|(|

professionals. That projection is hard to square with the Census Bureau's 2014 discovery that 74 percent of STEM majors in the United States work in non-STEM occupations.

A final unkillable claim among researchers is that science is broke, despite decades of bipartisan support for increases in science agency grant funding—the National Institutes of Health budget, for example, has risen from $11 billion to $40 billion in the last quarter century. The bottom line is that nobody in science ever says they have enough funding.

The Takeaway

Both the Covid-19 pandemic and the threat of climate change underline the need for science reporters to go beyond reporting discoveries to investigating decisions about science. Stories need relevance, insight, and originality to have any impact. The key to a revelatory story is to follow the science behind policies to show where things went wrong. Be on guard for claims from scientists of unfair shortfalls in students, funding, or prestige resulting from your reporting.

34
Artificial Intelligence

Matthew Hutson

Matthew Hutson is a contributing writer at the New Yorker *and a freelance science and technology writer whose work has appeared in* Science, Nature, Wired, *the* Atlantic, *and the* New York Times, *among other publications. He was an editor at* Psychology Today *and is the author of* The 7 Laws of Magical Thinking: How Irrational Beliefs Keep Us Happy, Healthy, and Sane *(Penguin). He has an ScB in cognitive neuroscience from Brown University and an SM in science writing from the Massachusetts Institute of Technology. He lives in New York City and enjoys fire dancing (poi) and playing bridge.*

Artificial intelligence (AI) infiltrates much of our lives in ways seen and unseen. It controls voice assistants like Siri and Alexa. It filters spam, curates news feeds, recommends movies, and provides search results. It diagnoses cancer, composes music, drives cars, and invents alloys. Behind the scenes, it even detects financial fraud, manages shipping logistics, and reviews legal documents.

Now is an exciting time to cover AI—as it will be for the foreseeable future. Once a niche area of computer science most visible in sci-fi movies like *2001* and *The Terminator*, its profile has blown up. Billions of dollars annually pour into research labs and startups, and conference attendance has grown exponentially, quadrupling over the last 5 years. The field is fast moving, each advance quickly topping the last. It's immediate, with clear applications in everyday life, and yet it's also philosophical, raising questions about what it means to think or be human.

I sometimes say I cover everything from silicon to societal impact, so if you choose this beat, you can expect to write about topics as varied as computer hardware, mathematical algorithms, practical applications, computing industry trends, the structure of academia, individual researchers, and the ethical implications of AI. Computer science can intimidate those who don't know Fortran from their foot, but it's just like any other scientific area: Take it slow, ask basic questions, and keep your ears perked.

Matthew Hutson, *Artificial Intelligence* In: *A Tactical Guide to Science Journalism.* Edited by: Deborah Blum, Ashley Smart, and Tom Zeller Jr., Oxford University Press. © Oxford University Press 2022. DOI: 10.1093/oso/9780197551509.003.0035

Know Your Basics

Experts disagree on what constitutes AI. One way to look at it is this: If a computer does something that would require intelligence in a person, it's AI.

Everything we have now is "narrow" AI, or good at some specific set of tasks. But many researchers aim for artificial "general" intelligence, AGI, which would match or exceed a person at most tasks. They debate how far away it is, whether it's possible, and even whether it's worth discussing. According to Will Knight, a senior writer at *Wired*, "It's like talking about faster-than-light travel."

Very roughly, the field of AI is split into two approaches. The first is symbolic AI, also called good old-fashioned AI (GOFAI), in which an algorithm is developed to follow a set of human-crafted rules: *If an animal in an image has four legs and whiskers, call it a cat.* The second is machine learning (ML), in which an algorithm—usually called a model—sees lots of data and adjusts itself until it finds its own rules, often impenetrable to a human.

In the last couple of decades, ML has begun to eclipse symbolic AI due to three factors: more data, more computing power, and new model architectures and training methods. Some researchers try to combine the best of symbolic AI and ML to reduce reliance on big data. A system might have some innate rules or structure to help it learn quickly with just a few examples. Children, after all, don't need to see a thousand zebras before they can recognize one; they need see only one, or even zero and the description "a horse with stripes."

Most ML falls into one of three categories. In *supervised learning*, the model trains by guessing and being told if it's right or wrong ("No, that's a dog"). In *unsupervised learning*, it looks for some underlying structure in raw data. It might say: "These images look similar to each other and these look similar to each other," without knowing that one cluster is of cats and one of dogs. In *reinforcement learning* (RL), the model learns to make sequences of decisions to achieve a goal. One controlling a robotic dog might eventually find the right set of movements to walk across the room. A few years ago, a model called AlphaGo beat the best human players of the Chinese board game Go in part by using RL.

Many types of models exist for each of these methods, the most headline-grabbing (though not necessarily the most common) being *neural networks*, which roughly mimic aspects of biological brains. These are complex mathematical formulas that pass bits of data between layers of small units of computation (neurons). When neural nets have several layers, it's called *deep learning*. I first experienced the magic of AI in a college course on neural network modeling. This was in the late 1990s, well before the deep learning revolution, with its feats of facial recognition and speech translation, so our systems were rudimentary. But just getting it to associate certain inputs with certain outputs, without my telling it how, was mesmerizing.

While neural networks have generated waves of enthusiasm followed by disappointment since the 1950s, deep learning really took off in 2012, after a model made a huge leap in performance in an image recognition competition. Another "AI winter"—a period when research interest and funding die off— seems less likely now that AI is finally generating boatloads of cash.

You don't need a technical background to cover AI. You can pick up the details as you go or write about issues surrounding the nuts and bolts of algorithms. If you do want to brush up, I recommend free or cheap online courses on AI and ML. *Medium* also has a lot of good explainers. For an over- view of the field, including technical performance, funding, and policy, check out the *AI Index Report*, updated annually.

Where to Find Stories

If you're used to covering other scientific fields, perhaps the first thing to know about computer science is that researchers publish mostly by submit- ting papers to conferences, not journals. Conferences conduct peer review— meaning two or more outside experts assess a paper's worthiness—and can be selective, accepting only 20 percent of submissions. The most important AI conference to have on your radar is NeurIPS (NIPS until 2018, changed for obvious reasons). I'll spare you the full conference titles; you can Google them, and they're generally known by their abbreviations anyway. Others good ones include: CVPR, ICML, ICLR ("I clear"), AAAI ("triple-A I"), IJCAI ("idge ick eye"), and ACL.

Conferences can overwhelm. Don't expect to absorb too much. People with PhDs in the field can stare with glazed eyes at work slightly outside their spe- cialties. Instead, ask people what they've seen that excited them. Or watch the invited speakers, typically senior scholars selected by committees to give longer and higher level talks, as opposed to authors giving short talks about their accepted papers. Or go to the socials and pick up gossip over drinks. And if you don't attend, browse the conference sites to see which papers have won awards, or ask the public relations (PR) officers to provide highlights.

Most papers appear on the preprint server arXiv ("archive") before submis- sion or acceptance. These are fair game for coverage. Some are never published beyond arXiv but still heavily influence the field. But beware. If you cover them, you'll need to do your own form of peer review by asking independent researchers to evaluate them. Many papers flood arXiv, so one way to tame the torrent is to use sites like Arxiv Sanity Preserver or Deep Learning Monitor, which track paper popularity. You can also follow researchers on Twitter and see what gets them all hot and tweety.

You may start receiving press releases from universities and companies. Sometimes companies will publish press releases or blog posts in place of a paper or well before a paper comes out. I usually avoid coverage unless there's something of substance I can ask another expert to examine.

Nevertheless, maintain good relationships with PR people at places like Google and Facebook, as it's often the only way you will get to talk to researchers at those companies—and tech companies publish a lot of conference papers.

Don't necessarily use any source suggested to you or quote anyone who remarks publicly on AI. Seek experts who have published on a relevant topic and be wary of those who call themselves "thought leaders" or are known for hot takes. Karen Hao, the senior AI editor at *MIT Technology Review*, notes a pet peeve. Beginning reporters "often just go to the loudest voice in the room," she says. "No one who understands anything about AI actually cares what Elon Musk says."

Ask Dumb Questions

Along with specific questions about a study, I have certain questions I typically ask researchers, in any field. To remember them, I came up with the acronym OCEAN. What are the **Origins** of this work—what inspired you, and what question did you want answered? Did you learn anything **Counterintuitive** or surprising? Are there any **Examples** of this phenomenon in the news or your personal life? (This makes more sense for psychology studies, which I also cover. For AI, you can substitute: What were the greatest **Engineering** challenges?) Do these findings have potential **Applications** in the near or distant future? Finally, what's **Next**—what are the work's limitations, and how do you hope to extend it?

At the end of an interview, I also tend to follow the acronym SEA. What other **Sources** should I contact? Are you **Excited** about any other projects? (This prompt can lead to future stories.) And is there anything else you want to **Add**?

For technical work, Hao asks researchers to walk her through the methodology, even if it's described in a paper. How did they collect, clean, and label the training data? (Where did the dog photo come from, and who decided it was a good example of a dog?) How did they train and test their model?

"Going through that pipeline is clarifying and can also produce really interesting details," she says.

Learn as much as you can before an interview. The conversation will be more productive, and researchers will appreciate your attention to detail; they will likely have seen work covered poorly by uninformed reporters. You should also fact-check your story, especially if the publication you're working for doesn't do it for you. Share as much of the technical descriptions with sources as you feel comfortable doing, as details are easy to bungle. You're not obliged to make any suggested changes of course—sometimes they improve accuracy at too great a cost to accessibility.

The general public knows very little about AI. So define terms and make explanations relatable. If a researcher asks me if I understand a concept, I'll often put my ego aside and say "Not fully." Because sometimes it turns out I actually misunderstand it, or the researcher provides a new perspective or a crystalizing quote.

In seeking lucidity, avoid anthropomorphizing—overattributing human-like traits. We're asking computers to perform tasks similar to those we perform, so we can easily slip into language or metaphors that overpromise AI's capabilities, saying it gets "creative" or "confused." Even so, some anthropomorphizing is inherent in the field, with established terms like artificial *intelligence*, machine *learning, neural* networks, computer *vision*, and natural language *understanding*, all in reference to the blind processing of ones and zeros.

A program's performance on a task, no matter how impressive, implies little about its performance on a separate task, or even the same one in slightly different conditions, revealing its utter lack of common sense. Knight says *his* biggest peeve is the hype around AI—this new shiny thing is better than all the others, so watch out puny humans!

Make It Personal

"AI researchers have a culture of abstracting things away from people," Hao says, but it's not all about algorithms when hunting down good AI stories.

You can cover the scientists creating the technology, the scholars discussing its uses and abuses, or the people directly affected by it. Hao, for example, has covered deepfakes—realistic images and videos fabricated by deep learning models. One early use was to place unsuspecting women's faces into porn. "I recently found a woman willing to speak on the record about her experience,"

Hao says. "So it's no longer a theoretical harm. It's a very visceral harm where someone's life was changed."

You might also cover the field from almost a sociological perspective—metascience coverage. I've done stories, for example, on researchers calling AI "alchemy"; problems with study replication; boycotts of a journal and a university; people and companies that annotate data so that machine learning algorithms can learn from it; the debate about whether peer review should consider technology's social impact; the difficulty of finding qualified reviewers to meet demand; the misleading use of performance benchmarks; and the denial of visas to conference participants from Africa.

These topics may seem like inside baseball, but they peel back the curtain obscuring how science operates, often revealing real drama. If you want color, just take to Twitter and watch the fighting words fly. Incidentally, this kind of story also provides a service to the field by highlighting crucial issues and stirring discussion.

In seeking people affected by AI, look for underrepresented communities. Not just potential consumers of some new app, but those surveilled by facial recognition or profiled by hiring algorithms. These may be women or people of color. I also try to include as least one woman or person of color as a source in each story. I defend this practice in two ways. First, it normalizes certain groups' presence in a field and industry dominated by White and Asian men. Second, it draws on the voices of people who may offer an unappreciated perspective due to their life histories.

Knight notes the importance of including the voices of experts and others outside of the United States, China, and Europe, especially in discussions of AI deployment and regulation. Decisions made about facial recognition, for example, can have repercussions around the world.

I also encourage more women and people of color to cover AI. "A lot of my favorite journalists and researchers in this space are women or people of color, who have really pushed the conversation forward in very important ways," Hao says. "This organically happens from people who have lived experiences that are different and make them more primed to be critical of some of the mainstream narratives. So yeah, I would just say join, please."

The Takeaway

Artificial intelligence is a fast-moving and multifaceted field, which makes it not only exciting but also tricky to cover. Make sure you get the science right—by educating yourself and working with expert sources—and don't overclaim AI's abilities.

Also, look beyond the science to who's creating and who's affected by the technology. Find case studies and the kinds of voices you don't normally see in a press release. Computers, even when using vacuum tubes, have never existed in a vacuum.

35

Cybersecurity and National Security

Kim Zetter

Kim Zetter is an award-winning investigative journalist who has covered cybersecurity and national security for more than a decade, initially for Wired, *where she wrote for 13 years, and more recently for the* New York Times Magazine, Politico, the Washington Post, Motherboard, *and* Yahoo News. *She has been voted one of the top 10 security journalists in the country by security professionals and her journalism peers and has broken numerous stories about the National Security Agency and Federal Bureau of Investigation surveillance, nation-state hacking, the Russian sabotage of Ukraine's power grid, and election security. She is a leading expert on the last, and in 2018 wrote a* New York Times Magazine *cover story on the crisis of election security. Zetter is also the author of* Countdown to Zero Day: Stuxnet and the Launch of the World's First Digital Weapon, *about a sophisticated virus/worm developed by the United States and Israel to covertly sabotage Iran's nuclear program.*

In 2006, a source reached out with what seemed like a wildly implausible tip. At the time, the United States was experiencing a record number of hacks, resulting in hundreds of millions of credit card and debit card numbers being stolen.

The source, David Thomas, told me he had orchestra seats to the hacking action and knew many of the players by their real names. He claimed he'd been working under cover for the Federal Bureau of Investigation (FBI) for 18 months running an illicit online forum where thieves traded hacking tools, stolen identities, and bank card numbers and planned their criminal ventures. There was one problem: Thomas was a professional liar. He was a middle-aged, down-on-his-luck grifter who had begun his criminal career writing bad checks and had ended it working as a mule for Russian hacker kingpins.

Thomas claimed the FBI had flipped him after an arrest in Seattle and installed him in an apartment from which he secretly ran the forum—all for the purpose of teaching the bureau the ropes of the underground hacking world. Thomas said the bureau did this while letting crimes plotted on the

Kim Zetter, *Cybersecurity and National Security* In: *A Tactical Guide to Science Journalism.* Edited by: Deborah Blum, Ashley Smart, and Tom Zeller Jr., Oxford University Press. © Oxford University Press 2022.
DOI: 10.1093/oso/9780197551509.003.0036

forum proceed without punishment. Victims and banks were losing millions, he said, while all the FBI did was take notes.

It was a bold claim, but there was no way the FBI would ever confirm it, or even acknowledge that Thomas had been working for them. This meant there was no reason for readers or an editor to believe the tale. Except that Thomas had receipts, as they say: hundreds of chat logs showing conversations between hackers on the forum as they discussed their schemes and the name of a public defender who witnessed the FBI jailhouse interview where Thomas was flipped.

I spent the next 6 months confirming every detail I could—tracking down victims who verified the criminal schemes plotted in the chat logs and getting the public defender to reluctantly acknowledge that the FBI and Thomas had made a deal. I even managed to track down the former manager of his Seattle apartment complex, who confirmed that the FBI had paid his rent. (Thomas said the FBI never paid him directly for his work.) The manager also revealed that an FBI handler had provided his cell phone number in case Thomas ever got out of hand.

By the time I completed my research, I was convinced Thomas was telling the truth. My editor at *Wired* was, too. To convince readers, however, we took the unusual step of publishing two dozen footnotes with the story, detailing everything I was able to confirm and noting what remained unverified.

I tell this story because it underscores the kinds of challenges you'll encounter when you cover cybersecurity and national security. You'll be dealing in part with sources from the hacking and intelligence worlds who are used to operating in the shadows, where deception and anonymity are the norm and where smoking-gun evidence is either classified and out of reach or easily manipulated. You'll need to be comfortable with those worlds and scrupulous with your reporting.

You'll also need to be comfortable with reporting on technology and have a grasp of how things—like encryption and back doors—work. That said, most reporters who cover cybersecurity don't have a technology or security background and had to build their knowledge from scratch. This should be a relief to anyone who feels intimidated by the complexity or vastness of the beat.

A Fast-Moving Beat

Cybersecurity has become one of the fastest growing and most critical beats in journalism due to its impact on nearly every aspect of our lives—banking, politics, elections, warfare, medical care, and critical infrastructure—and its implications for both the health of the economy and national security. The 2021 ransomware attacks that led to the Colonial Pipeline shutting down for nearly a week and caused hospitals to turn ambulances away highlight the increasing stakes at play. And revelations about Russian efforts to hack election infrastructure or about repressive regimes using surveillance software to hack the mobile phones of human rights activists and journalists are good reminders of how important security is for everyone, not just corporations and government.

In light of this, it's surprising that cybersecurity is a relatively new beat in mainstream media. It used to be primarily the domain of the tech press—*PC World*, *Wired*, and CNET, to name a few—but began to get wider press coverage in the early 2000s with the growth of e-commerce. As the public moved online to shop and do their banking, so did criminals, who found vast portals suddenly open to them for stealing credit card numbers and banking credentials. Even then, however, reporting on cybersecurity was just an add-on to other beats; journalists writing about consumer tech, politics, crime, health, and finance found themselves covering security and hacking incidents only when an incident touched their beat.

It was only after a series of high-profile cybersecurity lapses a few years ago, including the massive leak of data by former National Security Agency (NSA) contractor Edward Snowden and the unprecedented hack of the Democratic National Committee and state and local election offices in 2016, that media outlets finally began to realize that they needed security-savvy reporters who could grasp complex computer issues and hacking techniques and translate highly technical concepts into accessible language for nontechnical readers.

Today, cybersecurity is evolving so quickly that it can be difficult to grasp for reporters who cover only the occasional breaking story. Without knowledge of the hacking and security communities and how they operate, without familiarity with the terminology they use and the distinctions between different kinds of hacks, reporters can succumb to common pitfalls—such as hyping insignificant events and making hackers seem more sinister and powerful than they really are.

Reporting is doubly challenging because of the security community's aggressive criticism of reporters who get facts and details wrong: The smallest technical error or reporting oversight can undermine all the other work you do, and you may find valuable sources ignoring your calls and emails the next time you need them for a story.

Good Sourcing Is Essential

The best way to ensure that you don't make mistakes is to cultivate good sources. They alone won't guarantee good journalism if you don't also have a solid grasp of the topic and know how to interpret what your sources tell you. But knowing who to listen to and who to ignore is key.

One mistake reporters often make is believing sources who aren't worthy of trust—either because they're intent on deceiving reporters or because they don't have the expertise they claim to possess or have ulterior motives for speaking with reporters.

Wired fell victim to a source in 2015, for example, who claimed he knew the identity of Satoshi Nakamoto, the anonymous and elusive creator of Bitcoin whom numerous publications had unsuccessfully tried to unmask. The publication's story claimed it had "the strongest evidence yet of Satoshi Nakamoto's true identity"—an Australian academic and entrepreneur named Craig Wright. No one had ever suspected him of being Nakamoto, but he fit "the cryptocurrency creator's profile in nearly every detail," the publication wrote.

The evidence included emails that the source claimed came from Wright's email account, an encryption key that appeared to be associated with both Wright and Nakamoto, and a post on Wright's blog, published just months before Bitcoin launched, that claimed he would soon release an important cryptocurrency paper. A few months later, the persona known as Satoshi Nakamoto debuted a cryptocurrency white paper on a mailing list, and 2 months after this, Bitcoin launched.

The evidence was circumstantial, and *Wired* found inconsistencies in some of it, but they published a story anyway under the headline: "Bitcoin's Creator Satoshi Nakamoto Is Probably This Unknown Australian Genius." The piece included some caveats—acknowledging that the publication might have been targeted for an elaborate hoax—but the headline seemed to suggest the publication believed Wright was Nakamoto.

Almost immediately, however, readers and other journalists found clues suggesting Wright may have intentionally altered some of the evidence—possibly to make him appear to be Nakamoto. They also discovered that Wright had fabricated information about his business associations and academic credentials. Though Wright never actually told *Wired* he was Nakamoto, months later he started publicly claiming that he was, without providing evidence to support the claim. In April 2019, three and a half years after its story published, *Wired* added an editorial note to the piece saying the publication "no longer believes Wright is likely to be the creator of Bitcoin." The headline on the story changed as well. It now reads: "Is Bitcoin Creator This Unknown Australian Genius? Probably Not."

Tricky Sources

Not all sources are pushing bad agendas, but they can be tricky in other ways, too. Some mean well but inadvertently pass incorrect information to reporters because they don't sufficiently understand the topic at hand. This occurred with a 2020 story that claimed that the security firm FireEye had been hacked by Russian spies who tricked a FireEye employee into revealing his network password via a phishing campaign. The source for the story was an unidentified staffer on Capitol Hill who was evidently briefed on the intrusion but either misunderstood the information they received or was misinformed.

In truth, the hackers did obtain employee passwords, but only after they had already slipped inside FireEye's network by hacking a different company named SolarWinds. The hackers, using sophisticated methods, embedded a malicious back door in SolarWinds software that FireEye downloaded from the software maker's website. This allowed them to enter FireEye's network and steal employee passwords stored there, then use those passwords to burrow deeper into FireEye's network.

The distinction between the two versions of events might seem small—in both tellings the attackers stole FireEye credentials. But the reported version suggested that one of the top security firms in the world fell for one of the oldest and simplest tricks in the hacker playbook, while the facts revealed a much different story, in which the intruders used never-before-seen tricks to pull off a highly sophisticated supply chain hack.

This raises another important issue: When talking to reporters, sources won't always distinguish between the information they know firsthand and the information they gathered from others. They also won't necessarily distinguish between what they know for certain and what they only suspect. I've sometimes re-reported stories other journalists have published only to find, when speaking with the same sources, that the information wasn't as solid as the original story conveyed. To avoid spreading misinformation, it's always good to ask sources how they know what they know and what they know firsthand as opposed to what came from others.

Journalists should also be skeptical when sources seem to be spreading FUD—fear, uncertainty, and doubt. Stories that overhype a problem can unnecessarily frighten the public and even lead to bad policymaking.

At the same time, reporters should also be skeptical when sources who aren't security experts downplay a security issue. For example, election officials and voting machine vendors have told reporters for years that voting machines can't be hacked because they're never connected to the internet. But in truth some voting machines are connected to the internet, as I discovered while reporting a story published in 2019. And even computers that aren't connected to the internet can indeed be hacked—in 2007, the United States and Israel launched a sophisticated digital weapon called Stuxnet against systems in Iran that were enriching uranium for the country's nuclear program. The systems weren't connected to the internet, so the attackers got their malware onto them via a rogue insider, who carried it into the secure facility hidden on a USB stick.

Consult Trusted Guides

You can avoid many pitfalls and increase the accuracy of your reporting by consulting trusted sources who aren't necessarily quoted in your story but can guide you in knowing what's newsworthy or not and help explain complex concepts in simple language. Respected professionals who run their own security consulting firms are in a better position to speak freely than someone working for a large company, who either can't speak on the record or can only speak through the filter of their company's public relations (PR) team.

They can serve as a sounding board when you need to verify the accuracy of what other sources tell you or help you interpret technical documents; for example, if a security researcher or firm tells you about a serious vulnerability in Apple products that could make it easy for someone to hack millions of devices, and Apple disputes the claim, an independent source with expertise in Apple security can help verify the researcher's claim.

One way to find such sources is to follow the Twitter accounts of reporters who are expert in the field you're covering and see which accounts they follow or note the experts who are quoted in their stories. You should also try to attend hacker and security conferences like DefCon, BlackHat, RSA, and Shmoocon. There are hundreds of other local and international security conferences held around the world, with some focusing on niche areas, such as election security or industrial control systems used in critical infrastructure. They offer a great opportunity to meet and cultivate sources when their PR gatekeepers aren't present.

One prominent example of what can happen when you don't consult experts occurred in 2013 after Edward Snowden leaked sensitive NSA documents to reporters at the *Guardian* and the *Washington Post*. Many of the documents were highly technical PowerPoint presentations that had little context to help reporters interpret them. One presentation was about the NSA's PRISM program—a data-mining program for sifting through emails, documents, video, and other traffic collected from Google, Facebook, Microsoft, Yahoo, YouTube, and other companies. The presentation revealed, the reporters wrote, that the NSA had a direct pipeline into the servers of these companies—essentially a back door—that let NSA analysts remotely search the traffic of foreign targets as it passed through the servers of these companies. It was an explosive story that the companies vehemently denied. And it wasn't true.

Once the *Post* published the slide on which the story was based, it became clear to people with technical expertise that the reporters had misinterpreted it. The slide was actually contrasting two ways the NSA can collect the data—either by secretly tapping undersea cables located in foreign waters that transmit the traffic to the servers of these companies or by getting it "directly from the servers" of the companies. The reporters assumed the latter meant tapping into the servers the way the NSA tapped the undersea cables. But the slide was actually referring to data obtained directly from the companies via a court order as opposed to being grabbed in a more roundabout way from the undersea cables. (The *Post* hired a tech expert after this to help interpret the remaining Snowden documents.)

One case where outside experts were key in helping reporters evaluate the authenticity of what a source was saying occurred with Guccifer 2.0, the anonymous figure who claimed responsibility for hacking the Democratic National Committee in 2016. In correspondence with reporters, the hacker claimed to be Romanian, but Lorenzo Franceschi-Bicchierai, a reporter at *Vice-Motherboard*, consulted with native Romanian speakers who reviewed a transcript of his conversation with Guccifer 2.0 and found oddities in Guccifer 2.0's word choices, sentence structure, and use of diacritics in an informal chat—all suggesting he wasn't really Romanian. The U.S. government eventually attributed the DNC breach to hackers working for a Russian intelligence agency, and the accepted theory at this point is that Guccifer 2.0 was likely working with or for the agency as well.

The Takeaway

Much of what I've covered here is focused on sources—cultivating good ones and avoiding bad ones. I've focused on this because years of reporting have taught me that journalists are only as good as their sources, especially on such a complex and fast-shifting beat where a reporter's expertise often comes courtesy of good sources.

The best sources can help make your career and provide you with story tips and expertise for years. But putting your trust in one or more bad sources can lead you dangerously astray and ultimately undo everything you build.

PART V

METRICS, MODELS, AND MARKETING

36

New Models for Science Media

Thomas Lin

Thomas Lin is the founder and editor-in-chief of Quanta, *an award-winning, editorially independent magazine published by the Simons Foundation. He is the editor of the anthologies* Alice and Bob Meet the Wall of Fire *and* The Prime Number Conspiracy. *Previously, Lin was a digital editor at the* New York Times, *where he managed the online science and national news sections and wrote about science, tennis, and technology. His work has also appeared in the* New Yorker, Tennis, *and* Wired *and has been recognized by the White House News Photographers Association, New York Emmys, the American Society of Magazine Editors, and others. He has taught at City University of New York's journalism school and volunteers on the Council for the Advancement of Science Writing. He has interviewed Nobel laureates and Fields medalists but is most proud of having edited and mentored some of the most talented science and math writers working today.*

If you've read anything about the state of the print news industry over the past 20 years, you're familiar with this narrative: The internet's advertising potential and promise of fast, free, unlimited content advanced like a killer asteroid, and many news publications, caught flat-footed as dinosaurs, left a trail of fossilized remains. According to the Pew Research Center, newspaper circulation in the United States is at lows not seen since 1940. From 2008 to 2018, newspapers on average lost 47 percent of their staffs and 62 percent of their ad revenue, as Google and Facebook came to dominate online advertising. The Hussman School of Journalism and Media at the University of North Carolina at Chapel Hill reports that, since 2004, about a quarter of American newspapers have gone out of business. Many news magazines have also downsized or gone out of print.

Collapsing news budgets have had a downstream impact on newspaper science coverage and many science-specific print magazines. According to data compiled by the Shorenstein Center and Columbia Journalism Review, the 95 newspaper science sections that existed in 1989 dwindled to 19 by 2012.

Thomas Lin, *New Models for Science Media* In: *A Tactical Guide to Science Journalism.* Edited by: Deborah Blum, Ashley Smart, and Tom Zeller Jr., Oxford University Press. © Oxford University Press 2022. DOI: 10.1093/oso/9780197551509.003.0037

Though many ad hoc sections focused on Covid-19 appeared online in 2020 and many journalists had to learn how to cover a pandemic on the fly, only a handful of dedicated science sections remain today.

What's more, the economic impact of Covid-19 has accelerated the decline in newsroom employment. The *New York Times* reported in December 2020 that "37,000 employees of news media companies in the United States have been laid off, furloughed or had their pay reduced since the arrival of the coronavirus."

Given the state of affairs, you could be forgiven for wanting to steer clear of science journalism. But you'd miss out on joining a special profession at a special time. Let me explain.

Survival of the Nichest

Despite the decline of science news in newspapers and the recent struggles of some legacy science magazines, by at least one measure there has never been a better time to be a science journalist. Mirroring the transformation and fragmentation of the media landscape more broadly, there are now more flavors of science journalism than ever before. This means more options and more entry points for writers breaking into the profession.

In 2012, with the news industry still reeling from the 2008 financial crisis, I left the science desk of the *New York Times* to join the Simons Foundation, a private nonprofit dedicated to "advancing the frontiers of research in mathematics and the basic sciences." There, I proposed and founded *Quanta Magazine*. I was convinced that substantive, carefully reported and edited news about fundamental research was something a lot of people wanted, even needed. At the same time, I was inspired by the business models of trailblazing Pulitzer Prize–winning nonprofit outlets like ProPublica and *Inside Climate News*, whose news operations are funded by private foundations and donations from the public. The idea that editorially independent "philanthrojournalism" could potentially help fill the void left by cutbacks in traditional science news and even lead the way to a brighter future for science journalism was as encouraging as it was exhilarating.

At first, no one had heard of *Quanta*. Writers weren't quite sure how to pitch the magazine or if they wanted to write for us. Our low-budget website looked like an early-internet blog. There were encouraging signs of the impact our journalism was having, but it took a few years and a lot of elbow grease to get *Quanta* off the ground, to hire a tiny staff and find our footing in the media landscape, and then a few more years to make the case for the budget needed to design a beautiful new website and grow the staff from 3 to 6 to 12.

Now *Quanta* has become the magazine I envisioned in 2012. Many millions visit the site each year; we routinely publish incisive, accessible, newsworthy articles that attract hundreds of thousands of readers; and our work has been honored by many of the top industry awards, including a 2020 National Magazine Award for General Excellence. I credit early staffers for recognizing the potential and sticking with the magazine through its startup phase.

Even as science coverage has generally suffered in traditional media (with the notable exceptions of newspaper stalwarts like the *New York Times* and the *Guardian* and expansions in cultural and technology magazines like the *Atlantic* and *Wired*), we've seen a superbloom of digital startups that report on science, health, or the environment. *Quanta*, ProPublica, *InsideClimateNews*, *Stat*, *Undark*, *Knowable*, *Kaiser Health News*, *Gizmodo*, *FiveThirtyEight*, *Vox*, and *LiveScience* represent some of the diverse species populating this modern ecosystem. Each is positioned within its own niche, with its own mission and business model. Let's take a closer look and explore what this new landscape means for you.

On a Mission

Whatever other personal or professional goals may motivate us, all serious journalists are fundamentally driven by a desire to discover and report the truth in the public interest. It's our unwritten mission statement, our journalistic equivalent of the Hippocratic oath. The best new digital science publications share this broader mission, but—unlike larger general interest newsrooms—also hew to their own hyperfocused missions.

For example, *Quanta's* mission is to illuminate advances in fundamental physics, pure mathematics, basic biology, and theoretical computer science through our independent, public service journalism. Yes, this is nerdy as all get out, and no, it's often of no immediate utility. But this curiosity-driven, human enterprise will someday revolutionize the way we live. Rather than join other news organizations in covering health, medicine, technology, engineering, space exploration, environmental issues, or science policy (as interesting and important as this is), my small team can make a bigger overall impact on society by owning our niche and being disciplined about filling a specific unmet need.

It also helps me make the case for continued funding, as success in nonprofit public interest journalism is measured in the currency of impact. (At *Quanta*, we monitor reach and engagement, reader and expert feedback, public dialogue, subsequent media coverage, classroom usage, and other signs that our reporting is being put to good use.)

As another example, *STAT*, a commercial enterprise that's part of Boston Globe Media, is mining journalistic gold with its hard-hitting reporting on "health, medicine, life sciences and the fast-moving business of making medicines," as its website proclaims, and more recently with its award-winning coverage of the Covid-19 pandemic.

Rick Berke, executive editor and cofounder of *STAT*, told me that his publication started up in 2015 with the belief that Boston as an epicenter in life sciences and medicine wasn't being covered aggressively enough. "*STAT* is not an academic publication, not consumer, not trade. It's something different," he said, referring to their hybrid approach. In just a few short years, he and his team have built a robust base of individual and institutional subscribers, in addition to advertising, events, and other revenue streams—enough to support a staff of about 60.

These newer models may or may not work for you. Journalists who relish covering any and all kinds of news, and who are almost genetically programmed to change beats every couple of years, could feel limited by a small newsroom focused on a narrowly defined goal. But others, like the talented ProPublica reporter Caroline Chen, have their own clearly defined mission and are at their best when it aligns with their organization's. As a healthcare journalist who has always wanted to do pure investigative reporting, Caroline found in ProPublica the rare place where she could do just that. Before taking the job, though, the former Bloomberg reporter asked questions about ProPublica's business model and financial structure. She said she didn't fully appreciate the benefits of a nonprofit mission-driven newsroom until she started working there.

"Fundamentally, impact is our top metric," she told me. According to its website, ProPublica succeeds when its "investigative journalism does more than expose wrongdoing and injustice; we intend for it to spark real-world change."

"As time went on there," Chen said, "I realized ProPublica's mission looks like my mission as a journalist."

Getting to Know You

One great thing about today's wide-open science media landscape is that you can try before you buy. I encourage you to pitch and write for multiple publications before deciding which to regularly freelance for or which staff might be the best to join.

On the flip side, the zoo of options also means more publication-specific content types, editorial practices, style preferences, and business models to become familiar with. Do your homework. Talk to the editors and reporters at each place. Ask about the organization's editorial requirements and workflow, mission, audience, pay rates, syndication policy, and sources of funding. Ask about its level of journalistic independence and about its conflict-of-interest rules. No matter how small the outlet you're pitching, always write on-point, fully baked story proposals, be at your professional best, and meet deadlines.

The benefits of these new business models are many. In addition to the mission-driven environment, they often have smaller, more intimate newsrooms, which can translate to greater individual attention from your editor and art staff and a more personal and caring culture. They are usually nimbler, faster to implement new proposals, and more willing to experiment.

Then there's work-life balance. At Bloomberg, Chen told me, reporters and editors responded to emails in under 5 seconds. "It feels like you're on all the time." While some of that comes with working for a wire service, major daily newspapers like the *New York Times* also operate at a breakneck pace. At ProPublica, Chen said, "People sprint when they need to," but you can also enjoy a work lunch during which people check their phones just once.

There are potential disadvantages to joining a startup news outlet, too, though these vary widely from place to place. Small new publications tend to have limited reach and brand recognition. Getting your byline in a major traditional newspaper or magazine, on the other hand, gives you instant visibility. (Syndication can give you the best of both worlds; for example, *Quanta* regularly allows *Wired* and the *Atlantic* to reprint our stories.) Joining a small startup also means wearing more hats and being able to quickly shift gears, while sometimes being the only person on the team with a particular skill set. Some salaries can be grant funded or dependent on donations. Pay can vary—big commercial publications pay more in many cases, but not always, and nowadays health, retirement, and other fringe benefits are often less generous at publicly traded media companies.

With the rise of specialized science publications comes greater demand for journalists with specialized reporting skills and deeper subject area knowledge. I'm seeing more science graduate students switch to writing, often via the acclaimed science writing programs at New York University and University of California at Santa Cruz or the fantastic American Association for the Advancement of Science (AAAS) Mass Media fellowship, which places scientists in training inside real newsrooms to get real-world journalism experience.

But ultimately, editors like Berke and me are looking for science journalists who both have a command of their subject and are strong all-around news reporters. More than anything, Berke said, he looks for people who are "sharp, curious, aware, someone schooled in the publication, who knows what's out there, who can cite stories, who can point to things he or she did that no one else did."

In other words, don't worry if your experience and expertise are currently weighted toward science or news reporting. All of the best science writers started in one camp or the other but have learned to do both well.

Find Your Place

It used to be that journalists followed a few well-defined career pathways. Some would start from the smallest local papers and work their way to larger regional papers, with a select few high fliers breaking into top-tier newspapers and magazines. Others would get a graduate degree in journalism in the hopes of leaping straight into a medium-size regional newsroom or even a junior level or fact-checking role in a bigger newsroom. Still others would clerk in major newsrooms while pitching stories on the side. Now, with the advent of digital journalism and a proliferation of new newsroom job titles, every journalist you talk to has their own distinct path.

I could tell you about my early professional wanderings, but Katherine J. Wu's more recent experiences are likely to be more relevant. In 2018, while getting her PhD at Harvard, she was placed at *Smithsonian* magazine as an AAAS Mass Media Fellow. After receiving her doctorate, she spent a year as a staff writer for the website of the science television program *Nova*. As the only writer in that role, she wrote a lot but was "dying for mentorship," so she went back to writing for the *Smithsonian* and joined *Undark Magazine* as a part-time web producer.

Then she got into the *New York Times* newsroom fellowship program. Before the fellowship started, Wu had freelanced for *Undark*, *National Geographic*, *Popular Science*, and *Smithsonian*. But covering science news during the *Times* fellowship, she said, was a "critical experience for me. Breaking news did get me visibility and trained me to think on my feet a lot."

Wu didn't see herself as a newspaper person, however, and knew that she "liked being able to break down the news instead of break the news." So when the *Atlantic* offered her a staff writing job, she accepted it, drawn as she was to the smaller, more intimate science desk; the focus on craft, structure, and narrative; and the way she's "encouraged to write about whatever I'm obsessed with."

While there's no need for your career path to resemble mine, Berke's, Chen's, or Wu's, whatever kind of publication you start writing for, look for opportunities to learn from more experienced writers and editors and give yourself a chance to try different kinds of reporting before you make up your mind. Finding a newsroom that's a good match for you can help you produce your best journalism and enhance your overall career satisfaction.

The Takeaway

Whatever kind of science you want to write about, there's probably a publication that caters to an audience that wants to read about it. Get to know the new mission-oriented specialty outlets and learn how to successfully pitch them stories. It can be a great way to break into the field and gain experience. It might not offer the same reach and exposure as a legacy title, but if it's a good fit and your missions are well aligned, it could be more rewarding overall.

If you're thinking about pursuing a staff position in a traditional, startup, or nonprofit science newsroom, try before you buy. Read voraciously, dabble, intern, experiment with beats, write for different outlets, and talk to other science writers. Having strong specialized reporting skills in one discipline could serve you well at a specialty shop, but so will becoming a solid all-round news reporter.

Finally, find your niche. A smaller digital science outlet might be a stepping stone to a big brand name publication, or it could turn out to be the destination you were looking for all along.

37
Measuring Success
in Science Journalism

Kate Travis

Kate Travis, a freelance journalist, is the former digital director at Science News, *where she was a staffer from 2011 to 2021. She and her team there oversaw digital news production, video, multimedia, audience engagement, analytics, and social media. She started her career in science journalism as deputy news editor, and later news editor, of the* Journal of the National Cancer Institute. *She has also been an editor for* Science's *online career magazine,* Science Careers, *and for a website devoted to careers in translational research. She lives in Washington, D.C.*

You have an excellent idea for an article. You get an editor's green light, research the topic in depth, interview the best experts, and write a compelling story. It sails through edits, and after it's published, there's a moment of relief before a basic question comes to mind: Did my story succeed?

What defines success for a particular story will almost certainly vary from place to place—and sometimes story to story. Broadly speaking, what stories aim to achieve will almost always be defined by the goal and the mission of the publication. Is the publication aiming to serve a specialized audience or the general public? A loyal audience of people who fit a particular profile? Is the goal to drive subscriptions within that audience? Is the publication doing large-scale investigations aimed at changing public policy? Is it covering basic research and discovery, with a goal of informing and delighting its audience?

Up until the mid-1990s, success for print publications was largely measured by metrics like circulation and viewership. Those numbers drove advertising sales, the key metric of business success, and reader surveys offered checkpoints on how audiences perceived a publication's content.

"Success in the old newspaper days was filling a hole in the paper," quips Dan Vergano, a science reporter at BuzzFeed News who spent many years

Kate Travis, *Measuring Success in Science Journalism* In: *A Tactical Guide to Science Journalism.* Edited by: Deborah Blum, Ashley Smart, and Tom Zeller Jr., Oxford University Press. © Oxford University Press 2022.
DOI: 10.1093/oso/9780197551509.003.0038

at *USA Today*. "Success was also reader letters saying 'good job.'" And on a higher level, success was measured by impact, such as some government agency changing a policy or legislation as a result of good reporting.

Other measures of success included if you were first on a story or got an exclusive that prompted follow-on coverage by competitors, got mentioned in other media or in legal proceedings, or won awards.

Most of those indicators of success are still in play. But now, folded in, are scores of digital metrics that add a level of quantitative analysis to the mix.

Metrics for Success

Both quantitative measures made possible by the digital news era and the long-standing measures of success just discussed can be broadly categorized in four groups, as outlined in a 2016 Reuters Institute report:

> **Reach:** Reach metrics measure volume. Unique viewers, for example, measures how many people visit a website in a given time frame. News outlets that were born on the internet—sites like Vox, Live Science, BuzzFeed News, and Web MD—reach an average of 23.5 million monthly unique visitors, according to a Pew Research Center report on 2019 web traffic at 46 digital-native websites. Compare that to legacy print publications like *Science News*, *Popular Science*, and *Scientific American*, which report, respectively, 1.4 million, 3.6 million, and 10 million average monthly unique visitors.

Unique page views measure the number of sessions in which a particular article was viewed. Most newsrooms use unique page views as a metric in some way. Just like unique users, the "typical" number of page views for a news story will vary: A successful story might be a number over 200,000 or 20,000 or 2,000, depending on whether you're at a large outlet with an advanced digital revenue and distribution strategy, a small subscription-based legacy publication, or a small site with a very specific, targeted audience. Likewise, the lower boundary that signals a story didn't reach its intended audience could be 50,000 or 5,000 or 500 unique page views.

That variation illustrates why unique page views aren't a good metric for comparing across publications. And focusing solely on increasing reach metrics can be bad for business—and for quality journalism.

Engagement: Metrics such as session time, engaged time, or how far down a reader scrolled are measures of engagement—a measure of time spent or action taken with content. For video, those metrics may be watch time or completion rate. Other types of engagement metrics can include how many people shared an article to social media or by email, how many wrote letters, or how many links were clicked in a story.

Loyalty: Measures of loyalty can gauge how invested your readers are in you and your content. Loyalty is usually measured as return users, but can also include other activity, such as signing up for a newsletter.

Impact: Did the story make a difference? Examples could include public mentions, petition signed, or policies changed. It could also be newsletter shares or attention by journalists or figures of authority or influencers.

For any publication or story, success will probably be defined by a blend of these metrics. For example, in 2020 at *Science News*, we, like many publications, launched a coronavirus newsletter. We placed a high value on reader survey feedback, which helped us understand the loyalty of the readers and the impact the newsletter had for them. We also tracked typical newsletter metrics like the rates for opening, clicking, subscribing, and unsubscribing, which helped us understand reach and engagement. When considered together, all those metrics let us know whether the newsletter was hitting its mark.

Success Varies by Audience and Story

"What you care about should guide what you measure," says Liz Worthington of the American Press Institute. "So if you are in an advertising-only business model, then you probably still do care a lot about pageviews. But if you're trying to focus more on reader revenue and you're really focusing on getting more subscriptions, you care about a different set of metrics. And if you're focusing on subscribers and keeping those people happy, you care more about subscriber page views and are they coming back."

Worthington is the director of Metrics for News, a software tool that combines quantitative metrics to create a blended, numeric score that can be customized to individual newsrooms. At *Science News*, we started using Metrics for News in late 2019 to really dig into which stories different types of readers engaged with. To understand what stories entice loyal readers, for instance, the blended score combines page views and engagement metrics for only frequent website visitors. A category focused on increasing views and engagement by casual readers includes those same metrics but for first-time visitors.

It's no surprise that the articles with the highest scores for those two categories are often different. For example, in the first quarter of 2021, the highest scoring story among casual readers was about the *Perseverance* rover landing on Mars, while a story on the impact of Einstein's general theory of relativity over the last century topped the list among our most loyal readers. The latter story had a smaller audience, but it showed high readership and deep engagement among our most loyal readers. The rover landing story also accomplished its goal by attracting new readers.

At the journal *Nature*, reporters and editors focus on serving the publication's core audience of loyal subscribers. So the staff pays particular attention to metrics that can give insights into successful stories for that audience, says Lauren Wolf, *Nature*'s bureau chief for the Americas. She offered an example of a story she worked on about sexual harassment policies at institutions. "It didn't get a lot of clicks in the way that I thought it would, but it resonated with the core audience. Those are two very different things."

Stories with the potential to change policies or business practices often have different metrics for success than general interest stories. Victoria Jaggard, executive editor for science at *National Geographic*, offered an example from the magazine's wildlife crime desk. "Those stories are so deeply reported," she says. "They're kind of depressing lots of times."

As a result, common metrics like global unique page views aren't necessarily a good measure of success. Yet a feature story on wildlife tourism that exposed the pitfalls of wildlife tourism "drove some social media companies to change their policies about [people posting] photos with wild animals," she says. "And that's impact. And that has a lot of meaning for the journalists who do all of that work trying to bring a spotlight on these issues."

The key is to not get too obsessed with the dozens and dozens of metrics that can be studied. "As science journalists we're drawn to data, we love everything to be evidence-based," Jaggard says. We tend to look at the stories at the top of those lists as the only ones that readers care about. "You start to feel like that carries more weight than it probably should, in my opinion."

How Analytics Guide What We Do

The misperception that publications are only interested in clicks is damaging for journalism and has eroded public trust in the news media. It comes from early attempts to earn revenue from websites the same way newspapers

272 Metrics, Models, and Marketing

and magazines did in the years before. "Obsessing about pageviews made newsrooms make bad decisions," Worthington says.

Page views, global unique page views, uniques—whatever you call them—simply aren't a valuable metric on their own for newsrooms or journalists. "If you're a staffer, then all of a sudden hitting home runs becomes a bigger priority than serving your beat," Vergano says.

Figuring out how to interpret analytics data requires a nuanced approach. "As science journalists," Jaggard says, "we have a responsibility to interrogate that data and to really ask ourselves, what does it mean? Was it curated correctly? Is it being interpreted correctly? That's fundamental to who we are in our reporting structure, and I think it's got to be fundamental then to how we approach data-driven journalism."

A good place to start in finding ways to use data to make smart decisions is to look at what *isn't* working—what analytics can tell you about which stories aren't that interesting to your core audience, for example. At *Science News*, one of our early looks at the stories that didn't reach our median number of page views revealed a largely random list, but a few common themes emerged—including a cluster of stories about spacecraft launches.

Space and astronomy stories usually perform very well by any metric of success—reaching a core audience, a new audience, or surpassing the average page view threshold. But launch stories didn't. They were below average by nearly every measure. And given that so many general interest publications also write about spacecraft, satellite, and astronaut launches, we decided to stop reporting on them unless there was a specific, notable reason to do so. Using analytics to decide against stories is a powerful tool for allocating resources where they will best meet a publication's goals.

On the flip side, analytics can be used to make smart decisions about which stories you *do* choose to pursue. "We do not have nearly as many reporters and editors as a lot of other outlets do," *Nature*'s Wolf says. "We have to be super selective about what stories we invest in. We know we can't always be the first, so can we make the best? Is there still a second- or third-day story that we can do? Is it worth it?" Keeping a watchful eye on what stories land with a loyal audience keeps a newsroom grounded in the stories most important for its audience.

Digging into certain elements of stories themselves—visuals, headlines, or story length, for example—can also give newsrooms feedback on how to tell stories for particular audiences. Many large newsrooms have the technical infrastructure to test different visual treatments, headlines, and advertising placements to better understand how those elements can affect a story's reach or user engagement. But even smaller publications can benefit from looking at those elements among the pool of stories that did or did not succeed.

A lot of journalism boils down to investment—in time and resources—and analytics can help set priorities and provide valuable feedback on whether you're reaching the audience you most care about.

The Takeaway

Analytics are just one part of a larger puzzle in defining what a successful story is and in figuring out what stories your readers want from you. Audience metrics should inform journalism, but shouldn't dictate it. "Analytics is both your best friend and your worst enemy," says Jaggard.

Instinct should also be a factor: Is a particular story idea novel? Surprising? Important? Standard news values still apply when deciding what to cover and when evaluating success.

An organization's mission and goals are also factors when pitching, assigning, and writing stories, so they should be considered when evaluating success, too, whether that's covering a particular topic, reaching a specific audience, or having a specific kind of impact.

It all boils down to a couple of essential rules: know your goals and know your audience. Learn the best ways to measure success and how to use the data—and trust your gut—in making decisions about what stories to cover and how best to cover them.

38
Social Media in Science Journalism

Liz Neporent

Liz Neporent is the executive editor of social media and community at WebMD/
Medscape. *She is a former health reporter and social media director for* ABC
News National *and the author and coauthor of 18 health and wellness books,
including the bestseller* Fitness for Dummies, *now in its fourth edition. She also
serves on the emeritus board and is a national spokesperson for the American
Council on Exercise and the Hudson Valley Women's Health Initiative. Liz is a
lifelong "pathological" runner, having run 25 marathons, 6 ultramarathons and
countless races at a variety of distances. In fact, when she's not with her husband,
Jay, and daughter, Skylar, she can usually be found in the gym or on the road.
Follow her on Twitter @lizzyfit.*

In 2010 or thereabouts, I was passed over for a lucrative health blogging gig
in favor of someone who clearly had less experience and knowledge than me.
The reason? She had 6,000 Twitter followers compared to my 300 and was
more actively engaged with her users, the editor told me.

I won't judge whether or not that editor made a sound call, but it did get my attention. I decided to learn everything I could about social media. Ultimately, this
shaped my career path. I'm now the executive editor of social media for WebMD/
Medscape, the largest medical news publisher in the world. For the past 5 years,
I've been running the company's 50-plus accounts across six different social
media platforms. Funny how a few tweets can lead you down a different path.

Your job prospects (or security) may never depend on your ability to craft
a pithy 240-character snippet. Yet it's an undeniable fact that social media
is now an integral part of journalism. Employers want their journalists out
there spreading the word about their work, sussing out sources, and becoming
"influencers" in their specialties. Consider this: There are 6,000 tweets sent
out every second, according to the website Internet Live Stats. That's the speed
of news today. You can't afford to miss out.

In this chapter, I explain how you, as a science journalist, can make your
social media voice heard and listen closely to others. It's a nuts-and-bolts

Liz Neporent, *Social Media in Science Journalism* In: *A Tactical Guide to Science Journalism.* Edited by: Deborah Blum,
Ashley Smart, and Tom Zeller Jr., Oxford University Press. © Oxford University Press 2022.
DOI: 10.1093/oso/9780197551509.003.0039

beginner's guide to help you understand the platforms that will work best for you, to learn the rules, to practice the right etiquette, to measure success—and to stay out of trouble.

Choose Your Platform

Of all the social media sites, Twitter is where journalism lives. Facebook and Instagram may have more users, but that's not where news breaks or where users go to exchange ideas in real time. Most journalists, including those who work in science and medicine, consider Twitter their most valuable social network, according to a 2017 survey by the journalism site Muck Rack.

That said, plenty of journalism is thriving on other platforms. TikTok is great for tapping into younger consumers; and, because it has a lighter touch, you can be creative in your presentation. The *Washington Post*'s TikTok account has helped gain the attention of tween and teen fans who might not otherwise follow their site. Messaging apps like WhatsApp help you communicate with small but highly engaged groups of users who actively sign up to receive and share your information. If you're trying to build a Latin American or European following, WhatsApp can be a strong play.

You aren't confined to typing either. Turning the camera on yourself can be a powerful way to raise your profile. Every major social media network supports video. Consider going live for breaking news, explainers, and interviews. Use scripted material to create prerecorded video that can be edited and shared across multiple platforms. After Google, YouTube is the number 2 search engine on the web, so it's invaluable for both posting and sourcing. (Note: YouTube is owned by Google.)

Social video doesn't require a huge investment. Your phone's camera and some free or low-cost editing tools do a perfectly acceptable job because users don't really care if what they watch is high quality and slickly produced, as long as the content is informative and compelling.

With any platform, using all of its features helps you grab the most eyeballs. Tweets of up to 240 characters and a link are the main way users interact with Twitter, for example, but the site also supports images, video, lives, polls, and 24-hour disappearing quick takes known as Fleets. And Instagram has five areas of functionality: the main grid posts you see when you scroll through your feed; "Stories" that last up to 15 seconds and vaporize after 24 hours; reels for short, fun video clips that include sound; Instagram TV (IGTV) for videos longer than 60 seconds; and Instagram Live. (There is also a Shop feature, which is likely not relevant for most journalists.)

Table 38.1

Platform	Recommended Number of Posts	Types of Engagements	Best for
(Twitter)	1–10× daily 1–3 Fleets weekly	Likes, retweets, saves, comments	Breaking news, influencer reach outs, scientific discussions, trends
(Facebook)	1× daily 1–2× Stories weekly (or auto post Instagram stories)	Reactions, shares, saves, comments	Groups, following publications and media sources
(Instagram)	1 grid post daily 1–10 Stories daily 1–5 reels weekly 1 live monthly 1 IGTV weekly	Likes, regrams, saves, comments	Influencer reach outs, good images
(TikTok)	1–3× daily	Likes, shares, saves, comments	Younger audience, influencer reach outs, trends, memes
(YouTube)	At least 1× weekly and up to 1× daily	Up/down vote, comments, shares	Scientific discussions, trends, influencer reach outs
(LinkedIn)	1× update daily 1× blog weekly	Likes, shares, saves, comments	Identifying and contacting sources, groups

The Takeaway: Although Twitter is likely to be your primary social media vehicle as a journalist, I recommend experimenting with the other major social media sites to see if they work for you. Be patient. Learning to navigate the quirky ins and outs of each platform takes practice. Also, novel social media apps pop up all the time, so keep an eye out for innovative newcomers where you can establish yourself as an influencer from the get go (Table 38.1).

Interact, Listen Up

If you view social media as a place solely to pour out links promoting your own stories, you are leaving valuable assets on the table. Social media is just that—media that is social. It's a conversation, a two-way street, a place to build relationships. The more interactive, responsive, and receptive you are within the confines of professional boundaries, the more benefits you will reap.

Most platforms feature likes, shares, saves, and comments as their main forms of engagement. Each network presents engagement metrics with a slightly different twist. Facebook expands the idea of likes to allow for a range of reactions. A Twitter share is called a retweet. YouTube features the "gladiator"-inspired thumbs up or thumbs down to rate videos.

My personal belief is that some engagements are worth more than others. Comments have the highest value because they require the most user effort. Saves and shares also rank high because they suggest the reader thought your piece of content was meaningful enough to either file away for future review or spread it to their own network. A reader scrolling through her feed passively tapping hearts makes "likes" the cheapest form of engagement, in my opinion. They're nice to have, but how much does the reader really care? Any day my comment sections are humming with troll-free, productive conversations is a good day for me. That's why I often directly ask readers to comment within a post.

I also believe that overall engagement is a far better measure of success than follower count. It is better to have 500 highly engaged fans than 5,000 who barely pay attention to you. You want the *right* followers versus the *most* followers. A following that hangs on your every word, shares your work, and strikes up a conversation in the comments section makes you an influencer—that's social media slang for VIP.

However, I also believe that the most essential reason journalists need to be on social media is to take in information. A finely tuned social ear speeds up research and allows you to make connections.

Mine your social feeds for stories that are bubbling up even before they hit other news outlets. Spot influential voices, trends, and sources. When I was part of a major national news team earlier in my career, I once identified a source who lived across the street from the location of a breaking story by searching LinkedIn's geolocation tab. We were able to get on-the-scene interviews no one else had. Before the rise of social media, tracking down a source like that would have taken days. We were able to do it within minutes.

Of course, if you're sourcing and crowdsourcing through social media, proceed with caution. Make sure the accounts you're interacting with are authentic and are who they say they are. Keep an eye out for red flags and suspicious behavior. I recommend verifying any information you get from social media with backup sources. You want to avoid the embarrassment of reporting on something misleading, inaccurate, or, worst of all, a hoax.

Beyond sourcing and trends, I listen in to see how my own stories are doing. If I've got something that pops, I send it out again. If a post is a dud, I will sometimes repost it with different post language. Because stories speed by so fast in people's feeds, there's no reason you can't send something out more than once. I have some content that performs so well I've been putting it up on a regular basis for more than a year. And if I notice a competitor has a story that is "going viral," I will look to see if I have anything similar that can "steal" their traffic.

The Takeaway: Thanks to social media, journalists have unprecedented access to their audience, putting them closer than ever to trends, story ideas, and sources. Sifting through social content has become an important part of a journalist's job. There are plenty of tools and services available to help you refine your searches, but start with the built-in search engines available "natively" in most major social media sites. They do a pretty good job of helping you find what you're looking for.

All Hail the Hashtag

Hashtags are a coded language unique to social media. To the uninitiated, a word with a pound sign in front of it may seem cryptic and arbitrary. But the genius of the hashtag is that it allows you to instantly find, join, and catalog a specific conversation. Hashtags give you the ability to zero in on what you want and weed out what's irrelevant. Hashtagged posts can be aggregated in one place, often with related gems you might not have discovered otherwise.

Searching hashtags is a key strategy for a science journalist covering, say, a conference. For example, people who attend the American Academy of Neurology annual meeting use the hashtag #AANAM in all of their posts when talking about the meeting on Twitter. When you type that hashtag into a social platform's search box, a string of posts that include it in the copy instantly pops up, which you can sort chronologically or by popularity, making it easy to see what attendees are discussing. Most conferences add the year in either a two or four digit format after a string of identifying letters to help keep the posts from each year's meeting distinct and to create a convenient historical record.

News events, topics, and special interests are also frequently associated with a hashtag or hashtag set. I especially like tags such as #WomeninScience, #WhiteCoatsForBlackLives, and #MedTwitter because they connect groups of like-minded thinkers. If you look at tweets using only the #heartdisease hashtag you will find discussions dominated by consumers who have heart disease discussing topics important to them, which is great if that's what you're after. But if you are looking for the heart disease conversation among cardiology professionals, the hashtag #cardiotwitter will return far better results.

Each social platform has its own hashtag etiquette. Adding up to three hashtags per tweet on Twitter is fine, and up to five on a TikTok. On YouTube, you can get away with 10 tags, and on an Instagram you can max out your allotment of 30 hashtags per post (and up to 10 on the Stories feature) without appearing insane. For LinkedIn, one to three hashtags suffice. Facebookers eschew hashtags for the most part because they aren't indexed and to avoid looking like they did a cut and paste from Twitter.

The Takeaway: Hashtags are a valuable tool on most social media sites. They are essential for participating in focused conversations and refining searches. It can take some poking around to uncover the hashtags that serve your purpose but be patient and pay attention to which ones top "influencers" use. Check out services like Symplur.com that curate hashtags explicitly related to medicine and science.

Maintain Ethics and Integrity

Putting yourself out there on social media gives you an opportunity to share your unique voice in a way that might not necessarily come across in your formal writing. You get to show some spirit and maybe have a little fun. Done right, your social presence helps you build a following and attract a larger audience for your work, but a single tweet can derail your career.

A *New York Times* writer lost her job after tweeting that she had "chills" after watching Joe Biden's plane land as he arrived in the Washington area for his inauguration as president. While this may sound perfectly innocuous to us, the *Times*'s policy regarding journalists on social media clearly states that "they should avoid expressing views that go beyond what they would be allowed to say in the paper." The *Times* does afford a bit more wiggle room for their opinion writers, but in this case the journalist in question was a news reporter.

Most journalists who run into trouble on social media do so for posting something that is obviously far more over the line. Often it's because they get caught in a flame war of nasty words with a keyboard troll and blast out an angry tweet without thinking it through or because they attempt to say something amusing that comes off as offensive. I rarely send a post without parsing every word, editing it a few times, and then letting it sit in the dialogue box for a few moments before I hit send.

Check to see if your organization has a social media code of ethics in place. If so, review it carefully. If not, the Society of Professional Journalists has a good one, as does the *New York Times*. No matter what, I strongly advise sticking to some kind of guidelines.

The Takeaway: Before you send any post you should ask yourself one simple question: Does what I am about to say in a public forum that can never be truly deleted have the potential to undermine my credibility and integrity or that of my colleagues or organization? If the answer is yes, or even if it's maybe, don't publish it.

39

Building Trust and Navigating Mistrust

Apoorva Mandavilli

Apoorva Mandavilli is a reporter for the New York Times, *focusing on science and global health. She is the 2019 winner of the Victor Cohn Prize for Excellence in Medical Science Reporting. She is the founding editor-in-chief of Spectrum, an award-winning news site on autism science that grew an audience of millions. She led the team there for 13 years. She joined the* Times *in May 2020, after 2 years as a regular contributor, and reported exclusively on the coronavirus pandemic. She has won numerous awards for her writing. Her work has been published in the* Atlantic, Slate, *and the* New Yorker *online and in the anthology* Best American Science and Nature Writing.

The earth is flat. The 1969 moon landing was staged. Dinosaurs and humans walked the earth together. Vaccines cause autism. Climate change is a hoax.

Misinformation about science is as old as science itself. Whether motivated by religion, money, politics, or just plain fear, groups of people have always sought to dismiss and malign science and scientists. But the immense power of misinformation to destabilize and destroy lives has perhaps never been on display quite as starkly as during the coronavirus pandemic.

I spent 13 years as editor-in-chief of an autism magazine, so I thought I had a good grasp on spotting and correcting distorted facts. But reporting on the Covid-19 pandemic has stretched my skills to their limit. There are waves of false information to counter, some of it borne out of ignorance, some willfully planted to create confusion.

With social media, antivaccination propaganda, and even the White House fanning the flames, just about every aspect of the pandemic is fertile ground for science denialism: Masks have become a symbol of freedom; drugs like hydroxychloroquine and ivermectin are touted as miracle cures despite overwhelming evidence to the contrary; and some people claim the virus was overhyped as part of an elaborate scheme to strip people of their independence.

Apoorva Mandavilli, *Building Trust and Navigating Mistrust* In: *A Tactical Guide to Science Journalism.* Edited by: Deborah Blum, Ashley Smart, and Tom Zeller Jr., Oxford University Press. © Oxford University Press 2022.
DOI: 10.1093/oso/9780197551509.003.0040

There is every indication that this fracturing of public opinion will persist long after the pandemic ends. In this morass of mistrust, how is a science journalist to gain readers' confidence? Do you spend time refuting lies or stay focused on the truth? And when even vaunted institutions like the Centers for Disease Control and Prevention and the World Health Organization bend to political interference, whom do you trust?

Trust No One

This may seem paradoxical, but a crucial first step to giving readers information they can rely on is to exercise skepticism of everything and everyone.

Don't buy into the idea that scientists and journalists are supposed to work together as a "team." Scientists, like all people, have their agendas, biases, and conflicts of interest. Bring the same skepticism to evaluating science that a reporter who covers government brings to analyzing policies.

"Make sure that you aren't taking people and the materials that they're putting together—whether it's a journal article or a book or whatever—at face value," says Brooke Borel, a fact-checking expert and editor at *Undark Magazine*. That old adage—your information is only as good as your sources— holds particular weight when you're wading into a controversial or nuanced area of science.

Anyone from a for-profit organization, like a pharmaceutical company, is likely to be beholden to the financial bottom line. Experts from advocacy groups or nonprofits may have specific goals, so vet the group's stance on all relevant aspects, as well as their funding sources. Academic researchers may be less prone to institutional influence, but it is always a good idea to ask them who funds their research and whether they have any conflicts of interest.

Sometimes even simple vetting, like googling the person's name, can uncover red flags. You should also familiarize yourself with your sources' work, for example, by making sure they haven't published questionable theories. Verify their reputation and that their work has their peers' respect.

Once you're confident that your sources are solid, make sure you have a clear grasp of the overall field and where any one source's ideas fit in. That usually requires reading and talking to a broad swath of experts.

Every field has reliable founts of information. For biomedical science, that may be reviews in journals like *Cell*, *Nature*, or *Science*. With climate science, solid resources include reports from the Intergovernmental Panel on

Climate Change and national climate assessments, which are published every few years.

"I would just start by going to those super standard, trustworthy sources, and then talk directly to the authors of the individual chapters you're interested in," says Lisa Song, who reports on climate science for ProPublica. "Guard yourself with information."

Digging into an organization's activities can offer surprising insights. For example, an industry group may, in their public statements, blame consumers for not doing enough to preserve the environment. But it may at the same time be lobbying against the very things it says consumers should be doing, Song says: "That would be disinformation."

Acknowledge Controversies

One of the cardinal rules in journalism, and especially science journalism, is to avoid false equivalence. When writing about Covid-19, for example, you do not want to give the same weight to people who claim that it is no worse than the flu as to experts addressing risk factors for death from the disease.

On the other hand, it is honest and important to acknowledge that certain aspects of science are rooted in controversy. Depending on the story you are telling, it may in fact be a crucial part of the narrative.

As an example, an article that discusses vaccine hesitancy among parents would be remiss in not acknowledging the elephant in the room: the disproven theory that vaccines cause autism. Ignoring the controversy would encourage the reader to fill in the gaps, perhaps with misinformation that's rife in that field.

Instead, earn readers' trust by addressing the falsehood head on. In the case of autism, charting the origins of the theory from the discredited doctor Andrew Wakefield, revealing his conflicts of interest and the retraction of his research, would offer readers the context they need to understand how the fallacious idea came to be and why it is wrong.

Likewise, hydroxychloroquine rose to quick fame as a successful treatment for Covid-19, and was heavily promoted by the Trump administration. Research soon proved the drug to be not only unbeneficial but also potentially harmful. But that news came too late to change the perception in many people's minds.

Writing about hydroxychloroquine without acknowledging this checkered trajectory might leave readers with a sense that the reporter was either ignorant or deliberately hiding reports of the drug's effectiveness—and would run the risk of losing the readers' trust.

Context Is King

As a society, we have not done a good job of helping people understand that science is slow and iterative, rather than a series of dramatic results. That puts the onus on journalists to describe not only individual findings, but also where they fit into what was previously known—and what is still unknown. "Most of us are not writing stories that cover every aspect of a particular issue," Song says. "We are zooming in on some small thing."

It is when this context is missing that articles can overhype and mislead, for example declaring one day that butter is good for you, and the next that it is not. These seeming contradictions can erode readers' trust and give them the sense that science is capricious and mysterious.

During urgent crises like the pandemic, the pace of information and the pressure to get it out to readers can be all consuming, leaving little room to spell out the caveats. But with topics that are literally a matter of life and death, it's crucial for journalists to be as cautious as possible.

That begins with deciding whether to cover something at all. Research is increasingly appearing first through preprints—papers that are posted online before they are reviewed by independent scientists. Preprints have long been the norm in physics and mathematics, but biologists have been slow to adopt them, citing a fear of being scooped as well as the danger of bad medical information reaching the public.

But in the early months of the pandemic, these concerns had to be set aside. Peer review can take months or years, time no one has during an urgent crisis, so most findings were released and covered by journalists while still in preprint form. This change in practice seems likely to continue postpandemic. Covering preprints takes more vetting than for a paper that has at least passed through the critical lens of some scientists and journal editors.

Ask yourself: Does the finding make sense in the broader context of what we know? Do the scientists spell out caveats clearly, or do they hype their results? Does the team have a track record of careful work? I've found that it is also helpful to send the preprint to several scientists and have them analyze the paper—to design a small-scale peer review of my own. And once you've decided to cover a preprint, let readers know that the work is preliminary and that it has not yet been reviewed for publication.

It can be powerful to give readers a glimpse of your own process and progress. For example, when covering the 2016 fire at the Ghost Ship warehouse in Oakland, California, the *New York Times* shared regular updates on what its journalists uncovered during their reporting, the documents they obtained, and the interviews they conducted.

Don't Amplify Misinformation

Some areas of science—vaccines, pregnancy, climate change, genetically modified foods—are hotbeds of half-truths. But nearly anything that has the public's attention can be exploited by misinformation peddlers.

"People who are spreading this are opportunists," says Davey Alba, a technology reporter who covers online disinformation for the *New York Times*. "They will latch on to anything that's in the news cycle at the moment."

For example, soon after the pandemic began, some quacks seized the opportunity to push questionable treatments. Antivaccine groups also quickly mobilized, seeding rumors among vulnerable communities as early as the summer of 2020. By the time the Food and Drug Administration green-lit the vaccines in December, conspiracy theories had infiltrated the mainstream, claiming that the vaccines permanently alter DNA, for example, or that they make women infertile.

But in trying to refute the falsehood, do not give it oxygen, Alba warned. By its very nature, dousing a piece of viral falsehood usually happens after it has already caught fire. "But the fact checks do not go as viral as the original piece of misinformation," Alba says.

At the same time, it's important not to spotlight something that has not yet become widespread. This puts reporters who are trying to counter fallacies in a tough position, Alba says: "You don't want to do it too soon, but also, if you don't get in there soon enough, it's useless."

Tools such as Facebook's CrowdTangle can help spot viral misinformation early. To avoid giving the lies more attention, don't link to them in your article or in your social media posts. The same goes for photographs and other images, which can travel across the internet without the context you provide in your careful debunking.

With today's sophisticated programs, discerning manipulated photographs is tricky, and the perpetrators often seem to be one step ahead of the image experts. Old-fashioned detective work can sometimes succeed where sophisticated imaging programs do not. For example, try to gauge when the photo was taken. Do the weather, the backdrop of the town, or the clothes people are wearing offer clues to its accuracy?

Avoid circulating a photograph that has been doctored even if your goal is to point out the manipulation. If you do need to include it, take a screenshot and embed it in the article, but add a red slash across it to mark it as disinformation, Alba suggests.

Chasing down falsehoods with facts can often seem like a losing battle, Alba says. "It's inherently seductive to have something be conspiratorial, as if it's some secret wisdom that you have in your pocket," she says. "That's the thing that is so hard about misinformation."

Be Transparent and Accountable

The more open you are about the sources of your information and your process, the more readers will trust you. Most people have only a dim idea of how journalism works. And in an era when people suspect reporting to be "fake news," transparency can build credibility with readers.

"They don't see your notebook and they don't see the file on your desktop with all of the sourcing and all the papers and all the interviews you've done," Borel says. "They don't know that you've done your homework unless you show them."

It's impossible to describe the original sources for every fact, but sprinkle them in wherever you make big claims and at least link to the rest. Mention the respected institutions your sources are affiliated with or the prestigious journal where the research was published. And clearly delineate what is fact and when someone is offering an opinion.

Before you publish, make sure your article is airtight. Read through the story line by line and verify that you know where each piece of information comes from. This is tough to do on short deadlines but highlight anything that needs confirming as you write and build time to go back to it at the end.

When Song calls her sources to check facts, she checks not only the quote but also the context and premise of the story. "That's a great way to catch mistakes," she says. If your publication has rules against checking quotes, consult an independent source to confirm that you understood the nuances.

Give people on all sides of a contentious issue the opportunity to comment. Many publications write "no surprises" letters to sources that summarize the main thrust of the story, so they cannot complain afterward that they were not given a chance to respond.

But if, despite your best efforts, you do get something wrong, be accessible and accountable and promptly correct the errors. I respond to as many reader emails and tweets as I can. Your process may look different, but it's important to have some way for readers to reach you about your mistakes. Revisions often don't get seen as much as the original article, so make them easy to find and clearly explain what was wrong and what you changed.

The Takeaway

If you have written your story on a topic you chose with care, selected the best sources, added the right context, checked your facts, and corrected errors, you've done your part. If you get backlash despite all this—which can be inevitable when writing about certain topics—that's not your problem.

With misinformation in particular, it is important to remember: The mob is not always right. The more you trust yourself, the more your readers will trust you.

40
Marketing Your Stories

Jason Penchoff

Jason Penchoff is senior director of marketing and communications at Memorial Sloan Kettering Cancer Center and a former vice president of member marketing at Medscape. *Prior to his 7½ years at* Medscape, *Jason led audience development initiatives for* The Economist, *including social media, paid search, and search engine optimization (SEO). Additionally, he has experience launching a digital media agency, and he started his career at a public relations firm. He also led public information campaigns for the United States Agency for International Development in Ukraine and the World Bank in Uganda. As the son of a lifelong journalist, Jason grew up with ink on his fingers and a strong respect for hard-working journalists.*

Today, producing great journalism is only half the battle. You also need to know how to market your stories. The days of submitting your copy and just starting on the next story no longer exists for writers seeking to have impact. You also need to consider your audience, experiment with content types, build your own brand, and use channels and tools to expand audience reach.

That sounds like a lot, right? But it doesn't have to be so complex or done all at once. The best way to move forward when staring at a blank page is to start typing. Marketing your reporting is the same way. Take it one step at a time and continuously build on your knowledge.

And while you're at it, make it fun for yourself. In science journalism, you get to write about topics that interest you. So, become an expert on those topics. Seek out other experts and stay on top of the latest science. Find out how to connect the topics you cover to your audience and where your audience is engaging on those topics. Take advantage of the available tools to expand your audience reach.

Jason Penchoff, *Marketing Your Stories* In: *A Tactical Guide to Science Journalism.* Edited by: Deborah Blum, Ashley Smart, and Tom Zeller Jr., Oxford University Press. © Oxford University Press 2022. DOI: 10.1093/oso/9780197551509.003.0041

Scanners Versus Readers

Audiences get their news coverage on a variety of platforms and multiple screens. How do you catch their attention? The best way is to understand your audience. For most reporting, you probably have at least two online audience segments: scanners and readers.

Many people who consume news stories are scanners who will only spend 15 seconds "reading" a web article. Scanners will glance through the story and try to pick up the main points or watch part of a video while scrolling through their social feed.

Does that mean long-form journalism is dead? No. Fewer people may read from top to bottom of an article, but the context presented in the longer format will provide additional credibility, and a smaller but potentially more influential group will read through the whole piece.

Creating Digestible Content in Multiple Formats

To reach your scanner audience, you need to make your stories simple to consume and easily digestible. Even in long form, you need an attention-grabbing headline and scannable subheads that break up the story. You may even want to create a subsection for the top takeaway points of your piece (as in this guide).

Find ways to create visuals for your story. Not only photography, but also infographics and other elements to help visualize your reporting and key data. These graphics can be part of the story or leveraged separately with a link to the story on more image-oriented platforms such as Instagram and Facebook. Short videos can be put together and shared on YouTube, TikTok, or other video platforms that might be appropriate to the audience you are targeting.

Amplify your audience reach by providing additional commentary, interviews, or analysis in an audio clip linked to your story that can be picked up by radio and podcasts. It is becoming more common to include an audio clip, and it can add greater context, credibility, or color to your article and provide yet another format to reach audiences.

Also, think about the different channels on which your reporting is being consumed. On Twitter, assume most people will only see your headline and no more than 240 characters. So, create specific copy for the tweet, make your article punchy, and get to the main point quickly. If it is a video, make sure there is a transcript displayed at the bottom of the video so someone can read it without audio.

Making your content digestible for scanners, expanding reach with multiple content formats, and customizing your content for important channels will give you more opportunities to cut through the competition and reach your audience.

Build Your Brand

These days, it is important for journalists to cultivate their personal brand and share their expertise on the topics they cover. There are a growing number of platforms to raise your profile. Look for the opportunities that you find the most relevant and authentic. Then, find ways to connect to your audience on those platforms.

The most accessible place to begin is social media (see Chapter 38). Start following the people and experts in the topic areas you cover—scientists, academics, and fellow journalists. Share their posts and start curating the information most relevant for your audience. If there are conferences or major news in your area of interest, cover that, too. Comment on why the study or clinical trial resonates with you and how it provides new insights. Over time, colleagues and experts will start following you—growing your network, increasing your profile, and expanding your reach.

There are live and virtual speaking opportunities for journalists to share their expertise. Before taking on public speaking, start with practice. There are excellent and inexpensive public speaking classes to build your skills. I have seen even the most seasoned public speakers refresh by taking classes or media training. Then, find smaller venues to practice in before launching yourself into bigger public speaking platforms.

A good initial place to start is podcasts. Identify the top health and science podcasts that intersect with the topics you cover and pitch yourself as a guest or suggest a topic based on a story you have covered. There are a number of excellent health and science podcasts, including Gimlet's *Science Vs*, WNYC's *Science Friday*, and Shawn Stevenson's *The Model Health Show*. Many medical, academic, and nonprofit organizations also have podcasts that you can pitch yourself as a panelist/interviewee. You could also start your own podcast or create a topic area for a "Club" (audio conversation on a specific topic) on Clubhouse (a social media app for group audio discussion).

Getting on expert panels, live or virtual, is another way to raise your profile. Look at the associations, conferences, and organizations active in the area you cover and offer yourself as someone who can speak on those topics. Once you build your brand to a certain level, there are speakers' bureaus that can help match you with opportunities for paid speaking opportunities.

Building your brand can be daunting. Start where you feel most comfortable and develop skills as you go. Ultimately, you are trying to expand your reach and build an audience for your work. Journalists today not only create stories but also become experts in the areas they cover.

Think About Search Engine Optimization

Search engine optimization (SEO) has come a long way over the years. In the early days of SEO, many tried to use tips and tricks to beat search engines such as Google. Now, SEO is more about making it easier for search engines to find and rank relevant content to a person's search query. What does this mean for you?

Always write for your audience, not the search engines. Most of what a search engine is looking for is consistency and a clear theme so it can match your content to a user query. When you're editing your piece, put yourself in a user's shoes and ask: What is someone interested in this topic most likely searching for? Use tools like the free Chrome browser extensions What's My SERP and Text Optimizer to compare any synonyms that might have a stronger SEO value. Then you can decide if you want to optimize your story with keywords and improve your SEO rankings.

A key element of SEO is refining your content format and structure. Similar to writing for scanners, use heading tags and subtitles to break up larger chunks of text. Using keywords you identified as the most relevant to someone searching this topic in header tags will also provide an SEO boost. Bullets or a numbered list with the key takeaways and in-article images optimized with a file title and alt tag (HTML attribute to specify alternative text) are other enhancements that help search engines categorize the relevancy of your story with a user query.

Search engine optimization has evolved so that most of what search engines are looking for are also best practices for making your content digestible to different types of readers and scanners. For health and science coverage, Google performs a higher level of scrutiny on expert credentials and authority.

Google categorizes certain web pages as Your Money or Your Life (YMYL) content and holds it to a higher standard. YMYL content includes health information, science news, and medical advice. Google ranks YMYL content according to its Expertise, Authoritativeness, and Trust (EAT) ranking system. While there are no definitive rules on how to improve EAT for a web page in Google, some established best practices include showing that an article has been reviewed by an expert or that it has been fact checked; linking to the publication's editorial policy that outlines standards for publication; providing user feedback options; and including an author biography with relevant credentials.

Amplify Your Story With an Omnichannel Approach

For your bigger stories, you are going to want to broaden your audience reach as much as possible. If you work at a media company that has an audience development team, work with them to augment your reach through all the channels available. If you are self-publishing, or your organization does not have an audience development or marketing team, there are still ways to expand the reach of your reporting with target audiences.

First, start with understanding and segmenting the audience for your story. Are there certain audience profiles or affinities you are trying to reach? For instance, perhaps your story has a clinical angle that would be of particular interest to healthcare providers. Or the story will be most relevant for academic researchers. If you are covering a consumer topic, is there a way to segment the relevancy of the story by profile, geography, or interest area?

Armed with knowledge about your target audience, you can start to think about audience channels. Are you covering a topic that would interest other journalists? Use your network or identify media that might want to cite or interview you. Search PubMed for who is also writing on this topic and may be interested in your story. If you have written an important story, perhaps you pitch yourself to radio media or podcasters for interviews. Or, if you wrote a big story for a trade publication, perhaps you pitch it to a national newspaper recast with more of a consumer angle. Also, send your story to any bloggers, podcasters, or curated newsletters that might interview you or include a link to your story.

Another excellent channel to use is email. If your organization has email lists, make sure your reporting is included in the appropriate newsletters. For your most important stories, send an email to your network and share with them why this particular story means a lot to you. You can also create your own newsletter. Platforms such as Substack allow you to build your own, independent newsletter channel.

Sometimes, your story will include experts and organizations that can also help amplify your story—but they often need a prompt to share with their networks. If your story has particular relevance or includes a positive profile of an association or institution, ask their communications department to promote your content through their channels, newsletters, or media contacts. Many organizations have a podcast or content-marketing program; find out if they will highlight your content or interview you. Additionally, if you have a prominent expert profiled in your story, ask them to promote your story on their own network.

If you are writing for an organization, find out if they have paid search or paid social programs and, for your most important reporting, ask that those stories get promoted. Also, if appropriate and your organization has a mobile news app, ask that they promote with an app notification alert.

For your most important stories, look at all of the outreach opportunities available to you and create an omnichannel marketing plan to maximize audience reach.

Use Tools and Analytics

A key ingredient to marketing your stories is understanding engagement metrics (see Chapter 37). How many people are engaging with your reporting, and how many are commenting/liking/sharing your story?

Many news organizations have analytics software (e.g., Google Analytics, Omniture) to measure traffic. Site metrics that might help you understand engagement include the following: How much time are users spending with your content? Are they returning visitors or anonymous users? Are they reading on their phone or desktop, and how many pages are they consuming? Which channels are most effective in reaching your audience, and which sources of affiliate traffic are linking to your piece? Another popular metric is overall volume and quality of comments as a proxy for user sentiment and depth of engagement.

Off site, you can also look at engagement at the channel level. For newsletter traffic, how many people were sent an email (volume)? What is the Open Rate (did you use the right subject line) and Unique-Click-Thru-Rate (how many people clicked into a link in the newsletter)? How many overall clicks did the newsletter get? How many people opted out or unsubscribed to the newsletter? Social platforms will also give you analytics with different names for each action that generally fall into the categories of Views, Shares, Comments, Likes, and Followers/Fans.

If possible, create a dashboard with the key metrics you are measuring. This will allow you to measure engagement metrics over time, create benchmarks for success, and build a potential testing plan (e.g., A/B tests for newsletters, visuals for a social post) for continuous improvement. Also, talk with the organization you are writing for about the metrics they care most about so you can align organizational goals with engagement metrics that you care about as the content producer.

The Takeaway

Creating compelling news stories is only a piece of the work of journalism: Making sure key audiences engage with your reporting is becoming an important part of the job. There are more and more ways to identify and engage users just as there is more and more competition for eyeballs.

Ultimately, it is about creating the right stories for the right audience in the right format and delivering it at the right time. Test and measure different formats, tools, and channels to see which ones help you most effectively reach your audience.

PART VI
THE GLOBAL PICTURE

41

Narrative Reporting Abroad

Martin Enserink

Martin Enserink is international news editor at Science. *Before that, he was an online editor and staff writer at* Science's *headquarters in Washington, D.C., and spent 8 years in Paris as European correspondent and European news editor. Martin has reported about global health and science around the world. He is a three-time winner of the Communications Award of the American Society for Microbiology, each time with another* Science *colleague; his story about the eradication of yaws, reported from Papua New Guinea, won the 2019 Communications Award of the American Society of Tropical Medicine and Hygiene. He has been a mentor for several years for science journalists in Africa in a program run by the World Federation of Science Journalists and wrote an online course,* Covering Ebola, *with Helen Branswell. He also helped organize the 2015 and 2019 edition of the World Conference of Science Journalists, respectively as member and cochair of the Program Committee. He is currently based in Amsterdam.*

I woke up in the middle of the night with a headache, nauseated, and gasping for air. It took me a while to remember where I was. Earlier that day, I had arrived in La Rinconada, a gold-mining boom town in the Peruvian Andes. At 16,700 feet, it's the highest city in the world, and I was suffering from acute mountain sickness. It was very cold and a stench of human waste and rancid frying oil filled the streets. La Rinconada has no running water and almost no sanitation. Crime and alcohol abuse are rampant. "This may be the worst place I have ever visited," I texted a friend.

But I now think of the story I came home with, published in *Science* in 2019, as a highlight of my science writing career. It covered an unusual study into the long-term effects of oxygen deprivation on the human body and was told through the eyes of the researchers and one of the 55 subjects, a gold miner. In the process, I tried to capture what life in La Rinconada is like.

Reporting from far-flung places is one of the more adrenalin-pumping experiences you can have as a science journalist. You'll have adventures, see

Martin Enserink, *Narrative Reporting Abroad* In: *A Tactical Guide to Science Journalism*. Edited by: Deborah Blum, Ashley Smart, and Tom Zeller Jr., Oxford University Press. © Oxford University Press 2022. DOI: 10.1093/oso/9780197551509.003.0042 science in action in the real world, and meet interesting people. But it's also a chance to shine as a writer. You're not crossing oceans for a routine research story and a quote or two. These are the kinds of stories you will sweat over for weeks, crafting a narrative that transports your readers from their sofas and mobile phones to a new world. Stories with interesting characters, some drama, and perhaps conflict stories, which your editor will proudly submit for awards.

I have covered science and global health for *Science* for more than two decades and reported stories around the globe. It took time to learn how to craft them. I never received any writing training, and like many science journalists, I tend to focus on science, facts, and evidence. The human side of the story often comes second.

In 2008, I traveled to Tigray, in northern Ethiopia, for a story about the introduction of a new generation of antimalarial drugs. It was not a bad story, but reading it now, I didn't make the most of the trip. It led with a farmer named Fanta Dargie who was a community health worker, testing villagers for malaria and providing them with drugs if necessary. From there, the story quickly veered into the science, the drugs' history, and global malaria policies, losing sight of the Ethiopian countryside. I did not spend enough time on the ground with Dargie, who lost two children to malaria. His story provided "color" but feels flat to me today.

Over the years, I have sought to distill my experiences into more powerful narrative stories. Doing so requires thinking about what you want to tell and how best to tell it, as well as careful planning before your adventure begins. Here's what I've learned along the way.

Should You Go?

Not everybody appreciates travel, let alone long-haul flights or days away from home, and if you don't, going to the ends of the earth may not be for you. There are many great science journalists who primarily work from their desks, occasionally venturing out to a lab or a meeting. Even if you love travel, there are drawbacks. You'll have a larger CO_2 footprint for one—but I like to think it's better spent on a good, original story than on another annual conference or weekend getaway.

The expense is an obvious obstacle. If you're not fortunate to write for an outlet that has a travel budget, there may be other ways to fund your trip. Scour the internet for fellowships and small grants. The Pulitzer Center is an amazing resource, and the application for their travel grants is blissfully simple.

You may worry about becoming a "safari journalist," the type of Western reporter who, ignorant of the local culture, parachutes into an exotic country, checks in at a four-star hotel, and produces a superficial story. That's a risk, but you don't have to be that journalist. Read up on your destination, make sure you know the basics about politics and culture, and come with curiosity and humility. You could also team up with local journalists. In the late stages of the 2003 severe acute respiratory syndrome (SARS) epidemic, I traveled to Beijing to cover Chinese science's dismal response, together with Ding Yimin and Xiong Lei, two Chinese reporters who knew the local scientific community very well and helped me set up and conduct interviews. It resulted in a scoop: We found out that military researchers had evidence that SARS was caused by a new coronavirus early on, but were banned from talking about it.

Fortunately, science journalism is on the rise in Asia, Africa, and Latin America, and reporters there are increasingly covering stories in their regions, sometimes supported by grants, and they travel far and wide, too.

So yes: You should go, if you have the opportunity and enjoy the thrills of a trip. It will increase your skill set and broaden your horizons.

Planning the Trip

In 2004, not long after I became *Science*'s Paris correspondent, I flew to Agadir, a city in southern Morocco, to report on a locust plague that had engulfed West and Northwest Africa. Eager to get to work, I showed up at the local office of Morocco's National Center for Locust Control, with a letter from the center's director promising a close-up look at the fight against the voracious insects—or so I thought. An official took a look at my fax, handed it back to me, and said: "This letter gives you permission to do an interview at the headquarters in Rabat." That would be Morocco's capital, 500 kilometers to the north, far away from any locusts. Befuddled, I read the French text again, carefully. He was right.

Good planning is essential—including clear arrangements about who you're going to talk to and what you will see. It helps to think of your story as a movie with several scenes and one or more protagonists. Explain to your sources that you'd like to follow them in different situations: doing measurements in the field, engaging with the community, and holding meetings, for example. When I wrote a profile about a charismatic young Palestinian neuroscientist in 2016, we spent a few days at Al-Quds University, just outside Jerusalem in the West Bank, but we also visited his parents, who lived near Jenin, to have lunch and talk about his upbringing.

If you're lucky, you will see good and bad times, success and adversity, during your stay. That takes time, so don't make the visit too short. If you think you need 4 days for everything you want to do, budget a week. Things will go wrong: Flights will be delayed, offices closed, interviews canceled. Extra time gives you flexibility and allows you to make use of unexpected opportunities. If you don't need them, start writing in your hotel or do some sightseeing.

Try to set up interviews with other sources in advance as well—scientists, politicians, patients, activists—even if they're only tangentially related to the story, or not at all. Many will tell you things you can use or deepen your understanding. Confirm appointments clearly and get people's phone numbers. Check out in advance where to buy a local SIM card so you can stay connected.

On some trips, you may depend on your sources, directly or indirectly, for food, shelter, or safety—for instance, when you're joining an excursion to Antarctica or, as I did, visiting a malnutrition treatment center run by Doctors Without Borders in the Sahel. Make sure they understand that, despite this relationship, you are an independent journalist and will feel free to write what you see.

Keeping the budget reasonable is a skill. Learn how to shop for airline tickets; there are all sorts of ways to save. (If you're Dutch, like me, this comes naturally.) Travel, food, and accommodation can be inexpensive in low- and middle-income countries. Many foreigners on business in Addis Ababa stay at the Sheraton, a glittering island of luxury with $250 rooms in a poor city. On the internet, I found a friendly, clean guesthouse run by an Ethiopian-German couple that cost $25 a night.

Consider whether you will need a local collaborator, translator, driver, or fixer, and if so ask other journalists and your sources how best to arrange for one. This will depend greatly on the country you're visiting and your own skills and experience.

In Morocco, things worked out in the end despite my misunderstanding. After making some phone calls up the chain of command, the local director sent me on my way with an entomologist to see some truly biblical locust swarms that were devouring orchards a few hours south of Agadir—before two small yellow planes swooped in to blanket the area with insecticides. And after I returned to Paris, I made sure to perfect my French.

Getting What You Need

You've arrived at your destination. Now you need to make the best use of your precious time, collecting as much information and as many impressions as you can. For a rich story, you don't just want to interview people; you want to see things happen.

In 2014, a team of virologists allowed me to tag along for several days in Qatar as they were trying to piece together how Middle East respiratory syndrome (MERS), a new coronavirus disease, was spreading from camels to humans. We visited camel farms, a racetrack, a slaughterhouse, and the health ministry. Each visit provided a new scene in the story, sometimes with vivid, gory details. Long drives in the desert were a good time to talk about emerging diseases and the important role camels play in Qatari society.

At night in your hotel, go over your notes and think about what else you need or could use. Maybe there is another field site you can visit or someone else to interview. Take every opportunity to see and hear more. This takes stamina, especially when you're tired and jetlagged. I've done boring interviews at the end of very long days when all I wanted was a beer, a shower, and a bed. But sometimes they yielded a nice quote that made the story a bit better.

Snap lots of photos, even if you or your editor have hired a professional photographer, not only photos of people but also of labs, buildings, places, road signs, food, and animals. They will capture details that will refresh your memory and enliven your story. I record all of my interviews. I can't write very fast, and when I do, I can barely read my own notes; I also like having everything on tape in case of misunderstandings. Needless to say, upload everything to the cloud as soon as you can.

Keep an eye out for interesting people and try to develop a relationship. In La Rinconada, I met a gold miner named Ermilio Sucasaire, who was enrolled in the study of chronic mountain sickness. He not only was happy for me to accompany him as he underwent a battery of physiological tests over several days but also showed me his house, a corrugated metal shack, and the mine where he worked. We talked about his hardscrabble life, why he had come to the city, and his hopes for the study. After I left, we stayed in touch on WhatsApp, and I checked in for some updates as I was writing. Sucasaire became one of the story's protagonists.

Writing the Story

You're back home. Now comes the hard part: stitching it all together. Resist the temptation to procrastinate. Start transcribing notes and organizing your material right away, plan additional interviews you still need, and start writing. Delaying will make things harder; memories fade and new developments can make your story outdated.

Of course, you want to write a captivating yarn. If the trip went well, you have the material: interesting characters and an unusual locale, and it is hoped some drama. The challenge is to weave that into a narrative that also incorporates the science, history, and other background information that gives the story depth and intellectual interest.

Again, it helps to think cinematically. In 2018, I traveled to a remote island in Papua New Guinea for a story about Oriol Mitjà, a young Catalan physician–scientist who has breathed new life into a 60-year-old plan to eradicate yaws, a sadly neglected bacterial disease. During my visit, there was a moment when Mitjà stopped at a village and found a young boy, crying, with nasty ulcers on his legs. Mitjá sat down to inspect the boy and quiz his mother. It made for a nice opening scene because it helped introduce him as a character. Elsewhere, his team set out to provide antibiotics to an entire village, but that operation was chaotic and not well organized. That scene worked well in the middle of the story, as it highlighted his challenges. I ended the story with a sharply contrasting scene: Mitjà, back in Barcelona, surrounded by VIPs in a theater where he received an award. In between, I weaved in the science of yaws, the history of the forgotten eradication plan, and critics' doubts about reviving it.

It can be tempting to insert yourself into the story. Some trips can be both physically and emotionally draining, and you may feel the need to share. At *Science,* we rarely do so, although there are exceptions. When our contributor Kai Kupferschmidt reported a story about a gruesome 9-hour trek through the Liberian forest to find a missing contact of an Ebola patient in 2015, the first person narrative worked well to convey just how difficult contact tracing can be. But in general, I suggest restraint.

You'll have far too much material, and discarding anything can feel like a waste. But even well-written narrative stories can be too long. It's remarkable how a set of fresh, experienced eyes can help. If you're lucky, you'll have an editor who will help you select the most important scenes, identify gaps, quicken the story's pace, and smooth its flow.

Some of your darlings will get killed in the process. The draft of my MERS story started with a scene of two Dutch researchers sipping white wine in an outdoor hotel bar in Doha, discussing their research plans for the days ahead. "I'm not sure you need that," my editor said. Instead, we decided to take the reader straight to a camel market, with a virologist shoving a large cotton swab up a camel's behind. It was the right call.

The Takeaway

Going on far-flung reporting trips is one of the most exciting and challenging things you can do as a science journalist. With good planning, curiosity, and writing flair, you will open new worlds for your readers and produce some of the most rewarding stories of your career. Happy travels.

42
Reporting in Authoritarian Regimes

Richard Stone

Richard Stone is the senior science editor for HHMI Tangled Bank Studios, where he oversees science content for documentaries and other nonfiction productions and manages media partnerships. Prior to joining HHMI Tangled Bank Studios, Rich was the international news editor at Science *magazine, where his writing often featured datelines from challenging reporting environments such as Cuba, Iran, and North Korea. He has contributed to* Discover, Smithsonian, *and* National Geographic *and is the author of the nonfiction book* Mammoth: The Resurrection of an Ice Age Giant. *In his spare time, Rich enjoys writing science fiction screenplays. He would like to thank Martin Enserink, Mara Hvistendahl, Andrea Pitzer, and Dan Vergano for their wise input.*

It was after midnight on October 25, 2017, and I'd just passed through passport control at the international airport in Tehran, Iran. After a long day out in the field, I was anxious that I might drift off and miss my flight.

That turned out to be the least of my worries.

No sooner had I settled down near the gate, when two grim-faced men in suits came over and told me to follow them. I suspected they were with state security and didn't see any point in arguing. They led me down a flight of stairs into an area for transfer passengers and then into a side room that appeared to be a VIP lounge. It had large comfy chairs and a center table with bowls of sugar packets. As an American journalist, I was by no means a VIP guest. I was being sequestered for an interrogation.

Up to that point, I'd had a marvelous time during my reporting trip for *Science*. Iran's health ministry, which sponsored my reporting visa, had arranged a last-minute visit to an environmental disaster zone: the mostly dried-up Hamoun wetlands in southeastern Iran. I got a firsthand look at the ecological and human toll and had a memorable lunch of kebabs with local rangers in an abandoned village.

Driving back to the provincial capital, my local hosts and I were stopped at a checkpoint. When the soldier saw my U.S. passport, I was escorted to a

Richard Stone, *Reporting in Authoritarian Regimes* In: *A Tactical Guide to Science Journalism.* Edited by: Deborah Blum, Ashley Smart, and Tom Zeller Jr., Oxford University Press. © Oxford University Press 2022.
DOI: 10.1093/oso/9780197551509.003.0043

guard post and waited while the commander investigated how an American journalist had entered the province without a permit. After a tense half hour, the province's governor ordered my release. But the interlude evidently raised a red flag with state security. They were waiting for me in Tehran.

As I braced for the interrogation to begin, I had no idea if I'd make my flight or end up in jail.

Risks and Rewards

For many science journalists, a bad day is when an editor turns down your pitch or a researcher declines your interview request. For those of us who report in countries with authoritarian regimes, a bad day could bring far harsher consequences. Why take that kind of risk? Easy: There is the prospect of coming back with stories no one else is telling—storytelling that reveals the nature of regimes that routinely twist facts or flat-out lie, storytelling that's both a personal rush and a public service.

Now, I'm not suggesting that you drop what you're doing and hop on the next flight to Pyongyang, North Korea. Some reporting can be tackled from afar—and this chapter includes some thoughts about how to go about that. But in case you are intrigued about the prospect of venturing as a science journalist into hostile territory, I have tips about how, as J. R. R. Tolkien might put it, to get there and back again.

Dozens of authoritarian regimes and police states exist today, but for a science journalist some are especially intriguing because of their reliance on science as a pillar of society or means of staying in power. Reporting in these regions—countries like Cuba, Iran, North Korea, and Russia—often carries substantial personal risk, as some governments view any journalist not employed by the media they control as a threat that must be managed or even view you as a spy.

For a journalist, the chief risk, of course, is the loss of one's freedom. Incarceration is a paramount worry, but there are other infringements as well: constant monitoring (bugs in hotel rooms and tails when you're on the move); constraints on whom the government permits you to interview; and individuals you may endanger by quoting their names and ones who don't understand the peril of speaking frankly with a foreign journalist. And while reporting on science may not carry the same mortal risk of reporting in a war zone, there is always the possibility that something could go terribly wrong—the killings of Daniel Pearl in Pakistan in 2002 and James Foley in Syria in 2014 are just two grim reminders.

As a matter of good journalistic practice, you bone up on the subject matter and interviewee backgrounds as best you can in the time you have. When embarking on a reporting project in an authoritarian country, careful preparation not only makes you a better interviewer but also is essential for mitigating risks.

Getting Started

How do you even find science stories in authoritarian countries? You can hunt for hidden gems in scientific journals. You won't expect to find North Korea's nuclear physicists or cyberwarriors writing papers in the open literature. But in 2016, volcanologists there coauthored a high-profile scientific article on Mt. Paektu, a volcano on the border with China. After getting onto the Paektu story, I tagged along on a trip that British scientists took to North Korea to evaluate the volcano's risk of erupting.

You can also search for documents a restrictive government may have quietly posted on the internet. A clinical trial registry notice pointed to one of the biggest science scoops of 2019: a renegade scientist in China who produced genetically altered babies.

Your own government's documents can also tip you to stories. The impetus for my 2017 trip to Iran came from a U.S. National Academies report, which included a brief mention of a unique research effort. Iranian scientists were probing the chronic health problems of tens of thousands of survivors of chemical weapons attacks that occurred during the Iran–Iraq war in the 1980s. My initial reporting pointed to a heartbreaking story that could only be told well after interviewing the victims and scientists in person.

In most authoritarian countries, it's difficult and risky to report while on a tourist visa. Obtaining a journalist visa is the first order of business—and requires patience. After first making contact with North Korea's Academy of Sciences in 2002, it took 2 years to receive my visa. What's more, I was constrained to an itinerary negotiated in advance and could interview scientists only in the presence of a government minder.

In an authoritarian state with fewer restrictions on movement—Cuba, for example—a reporter usually must still set up interviews and make travel plans far ahead of time. But there will be greater opportunity for deviations, particularly ones that your hosts deem may advance their interests. Just remember you have to play by their rules, which might require time-consuming and costly hassles, such as obtaining a press pass issued by the regime.

On-the-Ground Reporting

Getting out of the office to report is energizing, and it's exhilarating when you step off the plane in a country whose government loathes objective journalism.

To navigate unfamiliar or hostile terrain, you'll almost surely need a local fixer. Such a person might handle logistics—confirming interviews, getting from Point A to Point B, changing money, and the like. Or, it could be someone with journalistic chops who can assist with the reporting or at least tell you where to get a decent meal on your budget.

How do you find a fixer? If you ask your scientific hosts directly, they may saddle you with a state intelligence agent or an incompetent brother-in-law in need of a payday. Local journalists or that country's foreign correspondents' club should have recommendations, as might outside scientists who have worked with your hosts.

In certain countries, the government will assign you a fixer—essentially, a minder—whose job is to facilitate approved interviews and site visits and, broadly speaking, to keep you from poking your nose in sensitive places. You'll have no choice but to work with this person, and you may need to pay a fee. But be considerate: If you break the rules, your minder likely will suffer repercussions. (In countries like China that don't foist a minder on you, you may still require approval from the international department of an institution—a research center, university, or government agency—to gain entry to conduct interviews.)

If you do hire a fixer, also remember that working with you can paint a target on that person's back, especially if you are reporting a sensitive story. Your fixer is more likely than you to be harassed or detained, especially if she or he is a citizen of the country.

In any authoritarian state, it's wise to assume that someone is aware of what you're doing at all times. Deviating from your itinerary could land you, your fixer, or the officials responsible for your visit in hot water. And be wary of shooting video or photos of your fixer. They may be reticent about asking you not to do so. But if you come under suspicion later, the imagery could increase their exposure should security services search your phone or computer.

You can mitigate risks by familiarizing yourself with a regime's red lines. The local foreign correspondents' club should have reports on journalists or fixers who have ended up in hot water and how that happened.

You should probably travel with a burner phone—ask foreign correspondents what they use. Google Fi, for example, is a cheap option for overseas coverage if set up in the United States, though check first to make sure coverage is available where you're headed (Russia and Venezuela? Yes.

Iran and North Korea? Sadly, no.). Share your itinerary and host contact details with your editor and agree on safety phrases that convey your level of your concern with a situation you may find yourself in. Bring extra cash—you invariably end up needing more than planned.

As a journalist, it should be possible—and in some instances necessary—to obtain a second passport. Several Middle Eastern nations, for example, bar entry to anyone with an Israeli immigration stamp in their passport. And currently, the State Department requires U.S. citizens wishing to travel to North Korea to apply for a special validation passport good for one trip. Journalism is one of only a few categories eligible for the special passport.

Have a plan for getting out if you suspect you've crossed a red line and before word of it travels very far up the chain of command. It should be a no-brainer for your editor to give advance approval to cover the cost of changing an air ticket to reach a safe haven before an aroused intelligence service gets on your case.

Handling Unwelcome Fallout

Kudos to you if you venture into hostile territory and make it back with a tale to tell. First, you'll need to vet your reporting as best you can. Whenever feasible, seek verification from independent experts. But some assertions are impossible to verify. During my 2004 reporting trip to North Korea, biomedical scientists there claimed to have successfully cloned rabbits from somatic cells. If true, they were the second team after a French group to achieve that feat. They showed me the pricey instrumentation they'd imported from Germany for the work, and they showed me what they said were the cloned rabbits, which were white with black splotches just like the original rabbits. But the findings were published only inside North Korea, and as far as I could ascertain no scientist outside the country had vetted them. I noted those limitations in my *Science* article.

Having assented to granting you a visa, an authoritarian regime is likely to pay close attention to what you publish. Take care to not expose your sources to retribution if you can help it. Individuals who criticize or embarrass authorities in your story may be detained, or worse. It's essential to confirm in the clearest possible terms whether sources are comfortable having their names in a story. Speak frankly with them about the potential consequences of going on the record.

Always take care with communications. Let sources as well as your fixer know that while encrypted programs like Signal or Telegram might appear to be a secure method of messaging, they should still set messages to delete after they are sent. Also, the name of a Signal group remains in your account even after you delete the group activity, so to be safe, give groups innocuous names.

Keep in mind that any reporting trip to an authoritarian regime might be your last one to that country—at least for a while. As a science journalist, you may be deemed less risky or troublesome than a political journalist. But never promise positive coverage or agree to such a request.

Don't pull punches in the hope of securing a visa for a subsequent trip. My 2004 story on science in North Korea ruffled feathers in Pyongyang, and officials told me that I would be banned from receiving another visa for at least 5 years. In fact, it would be 7 years before I could get back into the country. Then, in 2013, I wrote about North Korea's accelerating tuberculosis epidemic and again drew the regime's ire and endured additional years of rejected visa requests before undertaking my most recent reporting trip in 2018.

That was just a few months after my harrowing experience in Tehran. In the airport VIP room, my two interrogators were menacing and surly. They had taken my passport, and the clock was ticking closer to my flight's departure time. I'd been up since before dawn, and it was approaching 2:30 am. Finally, they left the room. Five minutes later, they were back and told me I needed to move quickly if I wanted to catch my flight. An amiable young interpreter escorted me to the gate just before the plane was about to push back.

Reporting on the Hamoun wetlands is one of my cherished memories as a journalist. But I had been naïve to the security situation on Iran's eastern frontier and put my hosts at the health ministry—and myself—in jeopardy. It was a profound relief when I later learned that the officials who arranged my trip did not suffer adverse consequences. Next time I travel to Iran, I'll ensure that all my permits are in order before arriving. Yes, there may well be a next time, as long as I come across an irresistible story and persuade my editor to send me back.

The Takeaway

Authoritarian regimes keep a tight lid on information and often purvey "alternative facts." Prep hard before a reporting trip to gather independently verified information, and be ready to work diligently after you return to vet the reporting you've gathered.

Avoid flying in under the radar: Reduce risks to yourself and to your sources by obtaining a journalist visa. You may end up burdened with a government minder, but in some countries, few scientists and officials would dare speak with you without such a chaperone.

And take care to protect your sources and fixers, especially if your story ends up criticizing or embarrassing a regime.

43
Collaborative Journalism Across Borders

Iván Carrillo

Iván Carrillo is a journalist, editor, and television and radio host based in Tepoztlán, Mexico, who specializes in coverage of science, health, and the environment. He is a member of the Global Community of Explorers of National Geographic Society and hosts the Ibero-American Scientific and Cultural News (NCC) broadcast in 20 countries in Latin America, as well as El Futuro del Planeta *(The Future of the Planet) on* EarthX TV *and the YouTube program* Atlas Aquática, *dedicated to the conservation of the oceans. He is cofounder and director of* Historias sin Fronteras *(Stories Without Borders), and has reported in Latin America, North America, and Europe. His work has appeared in* National Geographic Latin America, Undark magazine, *and* Newsweek en Español, *as well as* CNN en Español *and the Discovery Channel, among other outlets. He is also editor and coauthor of the 2019 book* The Present, Past and Future of Science as Seen From Journalism.

In the summer of 2019, the nationally acclaimed reporter Valeria Román of Argentina joined a team of Latin American science journalists who were experimenting with something new: a multicountry reporting project that they would call Historias sin Fronteras (Stories Without Borders). Such collaborative reporting was itself nothing new—the International Consortium of Investigative Journalists (ICIJ), for instance, had long connected journalists from around the world to coordinate on award-winning global investigations.

But a collaboration purely built by science journalists was something more unusual. And in Latin America—a vast region containing more than 40 nations stretching from Mexico to Chile with more than 600 million people speaking dozens of languages—it was not only a pioneering idea but also a daunting one. Román knew the research terrain of Argentina, where she had worked as a reporter for the country's largest newspaper, *Clarin*, and

Iván Carrillo, *Collaborative Journalism Across Borders* In: *A Tactical Guide to Science Journalism.* Edited by: Deborah Blum, Ashley Smart, and Tom Zeller Jr., Oxford University Press. © Oxford University Press 2022.
DOI: 10.1093/oso/9780197551509.003.0044

had coauthored a successful book on evolution. Outside its borders, though, remained many unanswered questions.

As her Argentine colleague, science journalist Federico Kukso, would point out: "For many, Latin America continues to be a 'terra ignota,' an exotic, unknown, unexplored region. Not only for the foreign gaze, but also for the region's own inhabitants, who due to the peculiarities of the international informative ecosystem tend to know more about what is happening in American laboratories than about the research carried out in neighboring countries."

The Historias sin Fronteras project, coordinated through the California nonprofit Inquire First and funded by the Howard Hughes Medical Institute, was meant to tackle exactly that problem. Teams of journalists were invited, in the summer of 2019, to submit proposals for multicountry projects. Román was part of a group that proposed looking at transgender science and the medical and cultural responses to the needs of transpeople across the region. She had proposed an investigation into that issue some years earlier after reading that transgender people in Latin America "had a lifespan of 30 to 32 years." Her local editors had not been interested—at all—in such an investigation.

But the question kept bothering her. So when the cross-border project was announced, she answered a call for proposals with the idea of an investigation into access to health services for transpeople in the region. And this time, she succeeded: The plan offered by Román and colleagues in Mexico, Costa Rica, and Venezuela was accepted as the first experimental reporting project for Historias sin Fronteras. She was thrilled. But it didn't take long for Román to become surprised—and then deeply dismayed—by the tales of transgender life she began discovering in her own country.

Her first interview was with a transgender activist in Argentina who had been kicked out of her family when she was a teenager and had become a sex worker to support herself. "She got her first formal job when she was 42 years old," Román recalled, adding that the investigation became a slap-in-the-face wake-up call. "I could not understand how I had not seen the terrible impact of discrimination and violence against transgender people before."

She and her fellow reporters communicated by email, Skype, and WhatsApp to build a picture that wove together the stories from each country. And Román discovered that the situation in other countries was even worse. Argentina, at least, had a gender identity law that authorized hormone treatments and surgeries for transgender patients at public hospitals. No such legislation existed in the other countries. "When I talked to the other reporters, we were all surprised because we were getting such sad stories every week."

The publication of their findings provided the first portrait of a health system failing transgender people across Latin America. It was such a compelling result that it propelled further growth of the cross-border project, which has since published multicountry stories on subjects ranging from Covid-19 to endangered dolphins and aggressive development in the Amazon rainforests.

As a cofounder, director, and editor of the project, I celebrate its success and as well as the teamwork by Román and her colleagues. I can also tell you that such collaborative projects in science journalism pose some real challenges and require careful management. But such group projects have also have real rewards: They allow us to build new networks of journalists and readers; they illuminate questions in science, health, and the environment that cannot be fully answered by taking a local or national approach; and they offer the promise of building stronger and alliances of science reporters around the world.

Using our project as an example, I explore some of the lessons we have learned that may prove useful for others considering such approaches.

The Power of Partnering

The specialized journalist in Latin America is a rara avis. Although there has been a recent boost in the training and promotion of science, health, and environmental journalists by international organizations like the American Association for the Advancement of Science, the Fundación Ealy Ortiz A. C. in Mexico, and the United Nations Educational, Scientific, and Cultural Organization, most media outlets in Latin America consider staff science reporters to be a luxury they cannot afford. News coverage tends to be fragmented and dominated by local corruption scandals, politics, violent events, celebrity gossip, and a lot of soccer, leaving little room for science or environmental stories.

In addition, while Latin American countries share many similarities, they also have profound historical, economic, and cultural differences. As a result, news coverage often suffers from "the omnipresent correspondent syndrome," in which a reporter from an international news agency or magazine essentially lumps together the whole of Latin America when covering events in Mexico City, Caracas, or Buenos Aires, without realizing that the three capitals are as distant culturally as geographically. The results, predictably, include inaccuracies, inadequate sourcing, reiteration of stereotypes, and story recycling.

Cross-border journalism can help address all these problems.

Teamwork can take many forms, but I've found that group workshops are a key first step in bringing together regional journalists and giving shape to ambitious cross-border projects. Virtually every workshop I've been involved in has ended with the first draft of a story in hand. My first experience was in 2019, when I brought a diverse group of 45 Latin American journalists together in Lausanne, Switzerland, at the World Conference of Science Journalists.

I asked them to cluster in subgroups of two to four journalists of different nationalities according to their thematic interests—biotechnology, energy, health, nature conservation, and so on—and spend 25 minutes discussing potential stories before choosing the best one for a 3-minute pitch to the other groups.

Even in that brief time, the teams came up with solid and original proposals, including coverage of bioenergy projects, entomophagy (eating insects) as an alternative food source, and pollution in factory zones in Chile, Peru, and Argentina.

The winning proposal, chosen by three international judges, received financial, editorial, and technical aid and eventually led to the groundbreaking story described previously on healthcare access in transgender communities in Argentina, Costa Rica, Mexico, and Venezuela.

That workshop became the catalyst for Historias sin Fronteras, which has thus far launched six multimedia projects and published five reports in Spanish, English, and Portuguese.

Our program builds on the success of other such collaborations. In any such discussion, it's important to acknowledge the pioneering example of the ICIJ, a U.S.-based nonprofit with its own reporting team and a network of investigative reporters from 100 countries and territories, as well as partnerships with over 100 media organizations. Their award-winning collaborative work includes the Pulitzer Prize–winning Panama Papers, which uncovered more than 11.5 million leaked financial and legal records exposing wrongdoing by secretive offshore companies. The investigation, published in 2016, featured cross-border reporting from scores of countries, and its impact is still being felt 5 years later in courts from Malta to Peru.

Shaping the Story

You've managed to gather a team of regional journalists to collaborate on an important health or science story. Where do you go from there?

At the outset, there will be logistical challenges to confront—many of the journalists will have never met in person and will need to report and communicate among one another via telephone, Skype, or Zoom. That's especially difficult when reporting in far-flung places, many of which are sparsely populated and difficult to reach.

As a result, careful strategizing among reporters, editors, multimedia producers, and photographers is essential.

In the case of our project on the Covid-19 pandemic, a team of four Central American reporters proposed an investigation into the role played by scientists, many of them censored, in government decisions to fight the spread Covid-19 in Guatemala, El Salvador, Nicaragua, and Costa Rica.

They began by identifying the basic elements of the story, including the potential protagonists (the censored scientific advisors); the conflict (scientific recommendations vs. economic or political interests) and the main questions that the story would seek to answer (Are the government responses based primarily on a scientific or political agenda? What is the methodology used for counting victims?).

Then we set up a production schedule for research, reporting, and editing; the development of infographics and other multimedia components; and the overall project design. Because many of the participants had other newsroom commitments, we also set up follow-up meetings and deadlines for partial delivery dates and final publication.

The editorial follow-up meetings, typically conducted via Zoom, were particularly useful for the participants to share on-the-ground reporting experiences and the project's production progress, while refining the coverage so it would appeal to both local and foreign audiences. They also offered a great opportunity for reallocating resources and sharing sources, as well as reporting from other outlets.

Often, these group meetings can yield unexpected insights. One conversation we had during the production of another story—about an endangered species of dolphin confined between two hydroelectric dams in the largest tributary of the Amazon—led us to visualize the narrative as a "prison story," one that became a powerful metaphor for how the local human population was essentially "trapped" in an energy production system that threatens the habitats of both species.

Risk and Reward

Despite democratic advances in many Latin American countries, journalists continue to be treated with suspicion and disdain by many authorities and civilians, especially in politically volatile nations and remote regions rarely visited by outsiders.

Sometimes the hostility manifests in a refusal to cooperate. Moises Martinez, a reporter with Nicaragua's *La Prensa*, who was part of our Covid-19 project, said it is usual for authoritarian governments to withhold information in order to control the message: "Silence is part of their political propaganda." And sometimes the hostility toward journalists manifests in more dangerous ways.

In 2020, Mexico was named by Reporters Without Borders as the most dangerous country for journalists. Other countries in the region, such as Honduras, Colombia, and Brazil, were also on the list. Though political journalists usually face the greatest personal danger, it also exists for science, health, and environmental journalists, who often go to territories where strangers are not welcome.

I once had to be escorted by police officers while visiting the Large Millimeter Telescope in Mexico in order to prevent potential assault or kidnapping. Another time, I had to hide my identity as a journalist from armed drug traffickers while visiting caves in Juxtlahuaca, Mexico, where bat conservation work was being done. The Historias sin Fronteras journalists working to cover a controversial highway project through the Amazon had to shorten their planned reporting trip after local residents warned to leave "because you can get out dead." And the residents worried—another important issue for journalists to consider—that reprisals might also come against people in the region who helped the reporters.

The risks aren't limited to Latin America. "The brutal reality of recent years is that journalists routinely risk their lives just for doing their jobs, even in countries once thought safe," according to the ICIJ in a statement on its website about its commitment to pursuing global accountability.

Even so, there is cause for optimism, as more people throughout Latin America come to appreciate the power and importance of regional science journalism—particularly health and environmental coverage—which can directly affect their lives. Using Historias sin Fronteras again as a case study, our recent project on the controversial use of genetically modified crops in the Andes Mountain region in Ecuador, Colombia, and Peru is one example of a serious issue that extends far beyond a single nation's boundaries.

And it's important to be aware of the risks, to be careful, and still work to tell stories that matter. Our report on the environmental threats posed by plans to connect Peru and Brazil by carving a new highway through the Amazon was published by *National Geographic Brazil*, giving even wider exposure to an important regional issue.

The Takeaway

When I spoke with Marcela Cantero, a journalist from Costa Rica, about the future of Historias sin Fronteras, she said: "It should be an active network of collaborators. Not just an isolated experience, but for this network that has already been generated in Central America to connect with the network that is functioning, for example, in Brazil and Bolivia and with the network that Costa Rica, Venezuela, Mexico, Argentina have created, and thus be able to work on issues regionally and in a constant manner."

Cross-border journalism is more than just a tool to publish regional stories—it can put a lens on often-neglected health and environmental issues that affect millions of people. It can connect not only stories but journalists. But it's important not only to practice it but to maintain and to continue building networks that support such ambitious science journalism.

44
Reporting in the Global South

Esther Nakkazi

Esther Nakkazi is a health and science reporter based in Kampala, Uganda, and the founder of the Health Journalists Network in Uganda, where she trains and mentors journalists in science reporting. She was a mentor and trainer in the Mentoring for Excellence project in science reporting for the World Federation of Science Journalists and contributes to media outlets around the world. She started her journalism career in 2000 at the EastAfrican, *a weekly regional newspaper based in Kenya. She is a graduate of Makerere University in Uganda and has a postgraduate degree from the Uganda Management Institute. She won a Knight Science Journalism Fellowship to spend the 2007–2008 year at Massachusetts Institute of Technology in Cambridge and was a journalist in residence at the Institute of Tropical Medicine in Antwerp in 2016. She is also a 2020 Chevening Africa Media Freedom Fellowship fellow.*

When I started out on the science beat more than a decade ago, reporting for the *EastAfrican* newspaper, it was a lonely affair. There was no science editor and limited training, and only a handful of journalists in Uganda were dedicated to the beat.

Science journalism in the Global South (broadly speaking, less economically developed countries) has made remarkable progress since then. A growing number of initiatives and programs have emerged to support aspiring journalists, as have new publications and platforms to bring their work to light. Stories about science, health, and the environment—stories that affect the lives and livelihoods of local audiences—are finally getting deeply reported coverage in Latin America, Africa, and other developing regions.

Many of these stories can only be accurately told from an insider's perspective. When research into human genomics took off around the turn of the century, for example, the African continent was largely overlooked—by both scientists, who left us out of studies that would lead to important breakthroughs in genetic medicine, and the journalists who covered

Esther Nakkazi, *Reporting in the Global South* In: *A Tactical Guide to Science Journalism.* Edited by: Deborah Blum, Ashley Smart, and Tom Zeller Jr., Oxford University Press. © Oxford University Press 2022. DOI: 10.1093/oso/9780197551509.003.0045 the stories. But a new generation of local reporters in places like Nigeria and Papua New Guinea has since shined a light on the importance of including Global South populations in genomic studies, doing so in a way that empowers and preserves the privacy of their communities.

We need more journalists who can tell these kinds of stories. The handful of senior science reporters in the field are already stretched thin. Few newsrooms in the Global South have dedicated science editors and dedicated resources for covering science.

The good news is that the Global South is rich with opportunities to break into science journalism. All you'll need is the right training and mentoring, a supportive network, unwavering journalistic standards, and a little determination.

Training

In many parts of the world, it's common to find science journalists who came into the field with a specialized background in science. Most reporters and editors in the Global South do not enjoy that privileged background and are trained instead in general journalism, communication, or the humanities.

That doesn't have to be a disadvantage. You don't need a science degree to report accurately, critically, and fairly, and most science editors I've spoken with agree that it is actually easier to work with science journalists who are nonscientists because they tend to have a better understanding of what would be interesting, surprising, or important for their readers.

But a journalist who is new to science reporting may need to do extra legwork to cultivate their ability to convey scientific ideas in reader-friendly language and to make those ideas both relevant and compelling for general audiences. Fortunately, a growing number of training programs are aimed at helping reporters in the Global South master these crucial skills.

A handful of these programs are housed in universities. Schools, including Makerere University in Uganda, Nasarawa State University in Nigeria, and Stellenbosch University in South Africa, are beginning to integrate science journalism into their undergraduate curricula. But science journalists in the Global South can also turn to newsroom-based courses for their training. For instance, SciDev.Net, a global news outlet based in London, has developed online training courses to teach basic reporting and writing skills, with a focus on researchers and reporters in Africa. Courses cover such topics as how to pitch ideas to science editors, how to report on foreign aid in science, and how to treat gender in science reporting.

Dr. William Tayeebwa, former head of the Department of Journalism and Communication at Makerere University, where he is currently a senior lecturer, says that while journalism training courses are important, they are only a start. After that, he says, "It is key to work on mentoring by pairing the students who are interested in reporting science with practicing science reporters."

Perhaps the biggest mentoring effort thus far for journalists in the Global South is the Science Journalism Cooperation Project (SjCOOP), started in 2006 by the World Federation of Science Journalists. As one of the original participants, I attended workshops on science reporting and later was among a group of journalists paired with a veteran science journalist in Kenya, Otula Owuor, the founder and managing editor of the digital magazine *Science Africa*.

As a result of that experience, I began my career as a science journalist and eventually became a trainer and mentor for others in the SjCOOP program. Many of the program's graduates have gone on to become senior reporters or editors in their regions as well as mentors for the next generation of science journalists and have helped establish or support science news outlets in the Global South.

Finding Sources

Journalists in the Global South, particularly those starting out, have to be aggressive and innovative to gain access to scientists and research. Most scientists in the developing world depend heavily on grants from nonprofit research organizations, and most governments in developing countries have very small budgets dedicated to financing research and publicizing science projects.

One crucial way a journalist can widen their network of contacts is to join journalist associations—local, regional, and even global—where they can track their colleagues' work, become aware of important research, and share reporting sources and guidance. The social media platforms of these groups can also be extremely helpful in identifying and locating potential sources. Journalists from both the Global South and Global North should consider such networks.

Abdullah Tsanni—a young science journalist and former early-career fellow with the Open Notebook, a nonprofit group that, among other things, supports science and health journalists in developing countries—uses a combination of methods to find sources. In addition to monitoring scientific papers and databases, his personal network, and social media, he is a member of the African Science Literacy Network, which helps him identify the best sources to approach. "What makes a good source for me is someone with a good history of research, [who] collaborates with other scientists, serves as a journal reviewer and is winning awards," he says.

Beyond that, some scientists will allow you to spend time with them in the laboratory. Another way to meet scientists is through media science cafés, usually held in informal settings. One example is the Health Journalists Network in Uganda, which I founded in 2011. We hold our meetings in an open space—usually in the shade of a large tree—to offer journalists and scientists a chance to meet and discuss research. It's inexpensive to run and gives scientists the freedom to speak openly without having to rely on elaborate formal presentations.

Similar informal programs include one in Kenya run by the Media for Environment, Science, Health, and Agriculture and ones in Zambia run by the Media Science Cafe and the Humanitarian Information Facilitation Center. All these efforts get funding from a New York advocacy group, AVAC, mainly to discuss HIV prevention, but they are also used to discuss other science-related topics.

In Indonesia, the Society of Indonesian Science Journalists (SISJ), operates a collaborative program with local scientists to discuss their work in the Bahasa language. "We team up with scientists and journalists to showcase their work, says Dyna Rochmyaningsih, a freelance science journalist and cofounder of SISJ. "By bringing them together in one group, we hope to bridge the gap."

One recent SISJ initiative trained 200 journalists across the archipelago to cover the Covid-19 pandemic. "We invited molecular biologists, virologists, as well as public health experts in our WhatsApp group," says Rochmyaningsih. "The experts share their expertise and the journalists share what's happening."

Doing the legwork of establishing relationships with local scientists helps give journalists in the Global South an advantage over international reporters who might attempt to cover local stories from afar, says Tsanni. "You are in a better position to tell the right story because you live there, understand the local context and nuances, and also have access to sources and places that nobody else has."

Staying True to Your Audience

In the Global South, journalists are constantly under pressure to tailor their narratives to international audiences. For instance, international research agencies and scientific institutions with deep pockets will often bring their work directly to reporters and sometimes pay journalists to attend workshops or "information days" where they explain their research.

"I think science journalists in the Global South must challenge this state of affairs and instead seek out the science done in their country that is of most interest to their readers—which might not be the research done by or funded by the biggest institutions, especially international institutions," says Linda Nordling, a freelance science journalist based in South Africa.

Journalists must question who is putting out information, and why, and make sure they aren't being used as public relations tools, she adds. In other words, it's essential for editors and journalists to understand their audiences— and to push to publish stories tailored for their communities.

That crucial task has been complicated by the rise of "donor journalism," reporting supported by philanthropists, foundations, and nonprofits, typically from the Global North. Donor journalism comes in different forms, ranging from support for individual reporters and story packages to the funding of entire news outlets.

Donor journalism can pose a risk to the editorial independence of newsrooms. Unlike in the Global North, where commercial advertising, organized events, and subscriptions bring in a substantial amount of revenue to support independent journalism, there is added pressure for newsrooms in the Global South to limit their work to those stories and issues that most resonate with their donors. Although the funders usually start off with good intentions, interference can happen in unconscious ways—through comments on the quality of articles, opinions on how things should be done during yearly review meetings, or recommendations on topics. Those subtle incursions can create gray areas and murky relationships.

"I think we cannot avoid donor journalism due to constrained budgets for the media industry," says Deborah-Fay Ndlovu, the communications manager at the African Academy of Sciences. "But we can find a way around it."

Donor journalism works best when both sides fully share their objectives and strategies and encourage transparency and mutual respect throughout the editorial process. It's important that journalists and newsrooms in the Global South not give in to donor pressure simply because they are obliged to get the funds. Regardless of who pays for our work, we must maintain a level of independence to do journalism of public interest and follow our own agenda.

The Role of Solutions Journalism

Science is often done with an eye toward solving social problems. From promising research into solar geoengineering to reduce the impact of climate change to novel technologies for addressing food insecurity, this brand of science has become especially important in the Global South and has helped spawn a form of reporting known as solutions journalism.

Solutions journalism uses rigorous evidence-based reporting to help solve social problems. "It encourages reporters to feature not just persons but a response to a problem, provide evidence of results, and discuss limitations," says Owuor. "It encourages communities that have similar problems to replicate what works."

Solutions journalism aimed at helping specific communities is gaining a foothold in a number of African newsrooms, including Science Africa and Nigeria Health Watch. These outlets are reporting on issues like the agriculture practices being used by communities to increase yields, the technology tools used to access mental health in low- and medium-income countries, and the use of drones to spread fertilizers or deliver medical supplies in hard-to-reach areas.

Young journalists who aspire to do solutions journalism can find a number of organizations that offer grants and training in these areas. Among the most comprehensive resources is the Solutions Journalism Network, a nonprofit cofounded by author Courtney Martin and *New York Times* journalists Tina Rosenburg and David Bornstein. It offers training and supports solutions-based reporting projects in communities around the world.

The Takeaway

As people increasingly recognize the multitude of ways that science impacts our lives, "science journalism is going to become even more important," says Ben Deighton, managing editor of SciDev.Net. And with "the pandemic, infectious diseases, and climate change, more journalists have become science journalists and realize the importance of reporting science accurately."

"The critical point is linking science to everyday lives," says Chaacha Mwita, a former journalist and now a communication and policy engagement consultant based in Nairobi, Kenya.

A diverse new generation of digital media publications, such as *Asian Scientist* and *Scientific Arab*, are answering that call. As more opportunities for reporting, mentoring, and training develop over time, the future looks promising for science journalism throughout the Global South.

Epilogue

Stay Curious, Question Everything

Tom Zeller Jr.

Tom Zeller Jr. is the cofounder and editor-in-chief of the award-winning digital science magazine Undark *(undark.org). Prior to the magazine's launch in 2016, Zeller spent two decades covering technology, energy policy, climate change, and the environment for a variety of national and international publications, including 12 years as a staff writer, editor, and visual journalist at* the New York Times. *During the 2013–2014 academic year, he was a Knight Science Journalism Fellow at the Massachusetts Institute of Technology.*

"Journalism," the storied news anchor Walter Cronkite once said, "is what we need to make democracy work." The crime novelist and attorney Andrew Vachss was even more emphatic: "Journalism is the protection between people and any sort of totalitarian rule. That's why my hero, admittedly a flawed one, is a journalist."

I happen to agree with these sentiments, and whether you're merely contemplating a career in science journalism, or already well on your way— ideally armed with the rich trove of tips and resources provided in the preceding chapters of this book—I hope you recognize both the honor and duty that come with upholding this mantle.

And having said that, I also hope you'll understand that while those and other noble journalistic ideals—fairness, skepticism, impartiality, detachment—are important to know and understand and appreciate, the truth is this job will have you regularly mired in confusion, frustration, moral ambiguity, and wrenching self-doubt.

You should embrace all of that, too, and in this concluding chapter I'll try to explain why.

Tom Zeller Jr., *Epilogue* In: *A Tactical Guide to Science Journalism.* Edited by: Deborah Blum, Ashley Smart, and Tom Zeller Jr., Oxford University Press. © Oxford University Press 2022. DOI: 10.1093/oso/9780197551509.003.0046

Journalism Is an Art, Not a Science

Many years ago, I was on assignment for a major magazine and tasked with accompanying a group of scientists working to develop a protected wildlife corridor stretching from the tip of Argentina to northern Mexico—an avenue nominally designed for the region's keystone predator, the jaguar. These researchers were working with local governments, regional scientists, farmers, loggers, private landowners, and Indigenous communities to identify tracts of land that the big cats were likely to favor in their movements. The goal was to link these parcels together and protect them, creating a vast highway for the jaguar's enduring genetic dispersal.

I met up with the research team, along with a photographer, in the dense rainforest of eastern Costa Rica, about 60 miles east of San Jose. They'd been working with an Indigenous community tucked deep into the canopy, an hour's walk from any road. The families there subsisted on hunting, small-plot agriculture, and modest livestock ownership. A single pig or goat, in this context, was precious, and the scientists were trying to understand the community's complicated relationship with jaguars, which were known to prey on their beasts. Shooting jaguars was the understandable reflex.

For the scientists, our presence meant a coveted opportunity to receive high-profile media coverage. And for the magazine, it seemed a ripe opportunity to capture photographs of jaguars in the wild—no easy feat, given their stealth and reclusive nature. But after several nights camped in the bush, and despite the numerous self-triggering cameras we'd set up along the many steep, undulating jungle trails, we were coming up empty.

Then someone brought up the pig. I don't remember how the notion first surfaced, but the scheme was simple: Buy a pig from one of the families, tie it to a stake overnight, train a camera on the scene, and collect mind-boggling images of a jaguar attacking and devouring prey.

I'm quite sure recollections would differ about how serious the idea initially was, but rhetorical or otherwise, the ensuing campfire discussions among and between the scientists and the journalists were animated and sometimes heated—with allies and opponents of the idea tracking in all directions. The researchers wondered aloud how such a thing could be justified on scientific grounds, while we journalists wrestled with our own convictions: Wouldn't the photo captions need to divulge to readers that the images were staged—and worse, captured in collusion with the scientists being covered? And everyone pondered, perhaps most poignantly, what such an undertaking would communicate to these families, who were already struggling with jaguar predation of their livestock.

On their face, these may not seem like difficult questions, but in the fullness of the conversation, some interesting nuances were raised. The purchase of a pig, for example, could represent an opportunity to significantly help this community financially, particularly if the price was commensurate with what our rich-world budgets could afford. One of the researchers volunteered that the photographs might yield new data on how a jaguar takes down its prey—and that it may be possible to compare the cat's markings with previous camera captures to identify a specific individual and thus learn more about its movements. And we all knew that, absent any photographs, there were good odds that the story would simply be killed, and an opportunity to highlight an ambitious and important conservation program—and the plight of these families—would have been wasted.

Were these compelling arguments?

Seek to Understand Others

That rainforest discussion, no doubt fueled by exhaustion and perhaps a bit of field whiskey, was something of a Potter's box—a device that some readers of this book may already find familiar. Developed by Ralph Potter Jr., the longtime professor of social ethics at Harvard Divinity School, the "box" is a set of four successive directives for navigating ethical dilemmas: (1) define the facts underlying the problem; (2) identify your relevant values; (3) delineate the principles that matter to you; and finally, (4) consider the loyalties that govern your decision-making.

Not surprisingly, the model has found its way into many journalism ethics courses, and while its structure and application can seem intuitive, and perhaps even obvious, I've found it instructive to occasionally run my own dilemmas through the Potter prism. In the case of the pig and the jaguar— and accepting the facts at hand—I might have said my values included things like truth, honesty, and fairness, while my more granular principles derived in large part from the code of ethics developed by the Society of Professional Journalists (https://www.spj.org/ethicscode.asp). My loyalties, meanwhile, were to not only the public, of course, but also my bosses, the publication, and more distantly, perhaps, the creditors to whom I owed money. (As with any other vocation, journalists do want to be paid for what we do.)

This is all rather straightforward, but the true value of Potter's device, I think, is the opportunity it offers to populate the box with the values, principles, and loyalties of other stakeholders. The scientists involved, for example, would surely surface terms like objectivity, accountability, and openness among their core values. Their operating principles might come from codes of ethics established by organizations relevant to their field, such as the Wildlife Society or the Society for Conservation Biology. And their allegiances would surely be to their institutions, as well as to their colleagues, their students, and of course, the Indigenous families whose land and cooperation were so central to their study goals.

As you embark on a career in journalism—and certainly in science journalism—it's important to embrace opportunities to both test the durability of your own convictions and thoughtfully consider how doing your job intersects with, countervails, or otherwise impacts the people you cover. Keeping a core set of principles in mind is helpful, to be sure, but in the day-to-day scrum of reporting, you'll quickly learn that, as with all of life, journalism can be—messy.

We shared meals with these scientists in the field. We drank with them. Was that ethical? Unbiased? Objective? Sometimes the researchers shared unguarded thoughts during idle moments around the fire or while preparing equipment. Was I interviewing them at that point? Did they think I was? At one point, a member of a local family—a middle-aged woman I'd interviewed earlier—seriously slashed her finger with a machete. It was not life threatening, but I had a first aid kit, so I tended to her wound. Had I crossed some line?

Imagine a less elaborate scenario—similar to one *New York Times* journalist James Glanz discusses in this very book—in which you find yourself holding embargoed research findings showing that the level of questionable chemicals present in a local aquifer are significantly higher than previously known. The researchers demand that you not publish news of the findings until a date and time of their choosing, and yet you know the information to be of great importance to your community of readers. You also know that your competitors with deeper pockets might get the story out faster than you. Do you honor the embargo? Why might you? Why might you not—and what values and principles and loyalties do you adhere to, or choose to violate, in making your decision?

Whose interests are you serving if you decide to honor or break the embargo? Your readers'? Your sources'? Your own?

These sorts of quandaries, large and small, will attend almost everything you do—and it can be easy to stop noticing them, wrestling with them, revisiting them, or discussing them. I would encourage you to try to keep attuned to these moments and choices and to keep your parsing of each decision alive in your mind. If nothing else, these sorts of reflective questions provide an alternative to the ponderous and often bottomless debate over journalistic "objectivity."

Avoid the Objectivity Trap

For some journalists—usually those who came up in the pre-internet era—an allegiance to objectivity can seem fundamental to the profession. It captures the idea, easily caricatured, that a reporter ought to be devoid of personal opinion and interested only in the dispassionate collection of facts and quotes from a judicious array of sources. All viewpoints should be evenly captured and readers left to weigh for themselves who or what is right or wrong. The legendary editor of the *Washington Post*, Leonard Downie, perhaps embodied this philosophy more than most: He thought journalists ought not even vote.

The counter to this, of course, is that there is no such thing as objectivity, and that journalism adheres to this principle at its peril. It can nudge even well-meaning journalists—particularly science journalists—to "balance" their stories on contentious issues with countervailing and often fringe viewpoints. But even more fundamentally, every act undertaken by a journalist, supporters of this line of thinking would say, is by definition subjective: whom to interview, what facts to include, where to go for documentation. No one could possibly interview *every* relevant stakeholder or document *every* relevant fact, after all, and the mere act of telling a story necessarily leaves some things out, while choosing others to include. These choices, critics argue, are very often driven by a journalist's conscious or unconscious biases.

More modern arguments hold—sometimes very convincingly—that aspiring to objectivity merely reinforces existing power dynamics. Progress comes through agitation, supporters of this line of thinking would argue, while adherence to journalistic "objectivity" has only served to dampen marginalized voices, prolong injustice, and otherwise slow and stall the pace of social change. Former *New York Times* writer Nikole Hannah-Jones, whose appointment as a tenured professor at the University of North Carolina became mired in charges that her work lacked objectivity, put this eloquently in a discussion with NPR:

> The harm is by pretending that the news we see is being led by objective arbiters of fact. It's just not based in reality. And I could give you a thousand examples from the perspective of communities of color, of marginalized communities, where what we're told is objective news is not. . . . When you see Black reporters, people presume they see our biases. White journalists, though, are often treated as neutral, as

if they are not going through the world in a racialized way or a genderized way, as if their class status and upbringing is not shaping their stories. But they are.

Examples of objectivity and balance gone wrong abound in science journalism: For years, journalists covering climate science, for example, sought out the viewpoints of scientific outliers who disputed the reality of global warming—all in the name of balance. This, many critics would argue, only helped to delay public understanding of this pressing problem. At the same time, other stakeholders argue that some journalists have become so invested in alerting readers to looming climatic changes that they too readily move beyond the evidence, painting future scenarios of civilizational collapse that are not well supported by science.

On Self-Interrogation

While they are still out there interviewing, digging up documents, and doing the work of reporting, those in that latter camp, I would argue, have stopped asking questions of the most important source there is: themselves.

That doesn't mean you shouldn't rely on the knowledge you accumulate as you cover a specific beat, or that you shouldn't trust your cultivated instincts as you press ever deeper into mastery of a certain subject, whether it's climate change, medicine, public health, technology, the environment, or any of the myriad other beats you'll be asked to cover. It also doesn't mean that you won't have opinions on these issues—perhaps even strong ones—that grow out of that accumulating knowledge base.

But if true objectivity is impossible, and cross-cutting interests between and among the people you cover are inevitable (and they are), then your best tack as a journalist—perhaps the only truly reliable one—is to continually check in with yourself. Is your certainty about something preventing you from asking a key question or preventing you from hearing the voices of people outside your intellectual comfort zone?

A good example of this comes from Brent Cunningham, former managing editor of the *Columbia Journalism Review*. Writing nearly 20 years ago, in a 7,000-word analysis of journalistic "objectivity," he described an early episode in which he wrestled with his own convictions:

In the early 1990s, I was a statehouse reporter for the Charleston Daily Mail in West Virginia. Every time a bill was introduced in the House to restrict access to abortion, the speaker, who was solidly pro-choice, sent the bill to the health committee, which was chaired by a woman who was also pro-choice. Of course, the bills never emerged from that committee. I was green and, yes, pro-choice, so it took a couple of years of witnessing this before it sunk in that—as the antiabortion activists had been telling me from day one—the committee was stacked with pro-choice votes and that this was how "liberal" leadership killed the abortion bills every year while appearing to let the legislative process run its course. Once I understood, I eagerly wrote that story, not only because I knew it would get me on page one, but also because such political maneuverings offended my reporter's sense of fairness. The bias, ultimately, was toward the story.

In the Knight Science Journalism Program's online science-editing handbook, former *Scientific American* news editor and veteran science journalist Robin Lloyd provided a very thoughtful sentiment that resonates with me when I read Cunningham's personal anecdote: "I constantly have to remind myself," Lloyd said, "to remain reasonably skeptical about everything."

I would only add that, as journalists, we should continually bring a bit of skepticism to ourselves and own beliefs and certainties, too.

The Takeaway

I'm happy to report that we did not tie a pig to a stake. We obviously could not make such a sacrifice in the name of good journalism any more than the researchers could justify doing so, at least under those circumstances, in the name of good science. And in the end, a piece about the jaguar corridor did eventually run, but much later, and without original photos of jaguars from that trip. Still, "The Night of the Pig," as I've come to call it, remains for me one of those pointed moments when the noble, almost sacred journalistic ideals and principles I'd long held aloft in my own mind—and that I do hope we've managed to impart over the course of the preceding chapters—came tumbling down to Earth.

That's another way of saying a textbook can only do so much. Your values and convictions as a science journalist will be constantly tested and challenged as you move through your career. Sometimes the answers will be plain. But even then, take some time to wonder why you think they are plain. It might not change your course, but the worst possible characteristic that a journalist can cultivate—and this is particularly true for journalists covering science—is a sense of absolute certainty in the rightness of your own vanishingly small view of things.

Resources

It would be impossible to squeeze everything there is to know about science journalism into a single volume. With that in mind, we have compiled a list of books, organizations, and online resources that should be valuable to any science journalist, whether you are looking to launch a career, connect with a science journalism community, or simply sharpen your existing skills.

Books

1. *Best American Science and Nature Writing.* An anthology series published annually since 2000, with each edition featuring a handpicked selection of the year's noteworthy science stories and essays.
2. *The Craft of Science Writing: Selections From the Open Notebook*, Siri Carpenter, Ed. (2020). A collection of articles on the craft of science writing, as told by contributors to the Open Notebook, an online community of science journalists and writers.
3. *KSJ Science Editing Handbook*, Knight Science Journalism Program at MIT. (2020). An all-digital resource (https://ksjhandbook.org) aimed at equipping editors with the skills and know-how to handle science stories.
4. *The Science Writers' Handbook: Everything You Need to Know to Pitch, Publish, and Prosper in the Digital Age*, Thomas Hayden, Michelle Nijhuis, Eds. (2013). A collection of insights from leading science writers on the ins and outs of the craft, with an emphasis on freelancing.
5. *News and Numbers: A Writer's Guide to Statistics*, Victor Cohn, Lewis Cope, Deborah Cohn Runkle. 3rd edition (2011). A classic text aimed at imparting skills to evaluate statistical claims in science, health, medicine, and politics.
6. *A Field Guide for Science Writers: The Official Guide of the National Association of Science Writers*, Deborah Blum, Mary Knudson, Robin Marantz Henig, Eds. 2nd edition (2005). A collection of essays combining practical, how-to advice with thoughtful discussions about the fundamental challenges of science journalism.

7. *Ideas Into Words: Mastering the Craft of Science Writing,* Elise Hancock (2003). A slim but insightful volume that covers the essentials of crafting a science story, from finding stories and interviewing to writing and revising drafts.

Organizations

1. **Association of Health Care Journalists** (https://healthjournalism.org). A nonprofit organization with more than 1,500 members, aimed at improving the quality, accuracy, and visibility of healthcare reporting.
2. **Council for the Advancement of Science Writing** (https://casw.org). An organization that seeks to improve the quantity and quality of science news, with a special focus on the quality, diversity, and sustainability of science journalism.
3. **European Federation for Science Journalism** (https://efsj.eu). A nonprofit that organizes meetings and conferences, sets up cross-border investigative reporting grants, and runs awards, among other activities.
4. **National Association of Science Writers** (https://www.nasw.org). A U.S.-based community of more than 2,500 journalists, authors, editors, producers, public information officers, and students who cover science and technology.
5. **Society of Environmental Journalists** (https://www.sej.org). A North American membership association aimed at improving journalism and advancing public understanding of environmental issues.
6. **World Federation of Science Journalists** (https://wfsj.org). A nongovernmental organization that represents more than 60 science journalists' associations worldwide and organizes the biennial World Conference of Science Journalists.

Online Resources

1. Council for the Advancement of Science Writing, *Resources* (https://casw.org/science-journalism/resources/). Contains links to resources on craft, career building, and professional development in science journalism.
2. Knight Science Journalism at MIT, *Being a Science Journalist* (https://ksj.mit.edu/resource/being-a-science-journalist). A compilation of

resources that includes links to academic science writing programs, internships, and fellowship opportunities.

3. National Association of Science Writers, *Writer Resources* (https://www. nasw.org/writer-resources). A collection of resources on freelancing, teaching science writing, getting started in science, and other topics, including some members-only portals.

4. Society of Environmental Journalists, *Resources* (https://www.sej.org/ library/main). A one-stop shop for an array of resources and journalism guides, including a "Climate Change Resource Guide" and a "Journalists' Guide to Energy & Environment."

5. The Open Notebook, *Getting Started in Science Journalism* (https:// www.theopennotebook.com/getting-started-in-science-journalism). An extensive resource page for beginning science journalists, compiled from essays and articles published in the Open Notebook.

6. World Federation of Science Journalists, *Resources* (https://wfsj.org/ resources/). A resource page containing information about science journalism webinars, Covid-19 reporting resources, reporting grants, and other initiatives.

Index

For the benefit of digital users, indexed terms that span two pages (e.g., 52–53) may, on occasion, appear on only one of those pages.

Tables are indicated by *t* following the page number

layered narratives, 57–58
legwork, 47–49
Levy, Steven, 195, 197
limitations of studies, 24
Lin, Thomas, 261
LinkedIn, 17, 115, 199
logos, 91
"Lost Mothers" project, 165–66
loyalty metrics, 270
loyalty tests, 115

machine learning, 82–83, 177, 245
magazine journalism
  background, 94–95
  goal clarity, 99–100
  organization, 96
  structure, 97
  tension in, 95–96
  visuals, 98
  voice, 98–99
magazine model, 36–38
Maines, John, 120
Makerere University, 319, 320
Malloy, Mike, 52, 58
Mandavilli, Apoorva, 16, 164, 281
Mannermaa, Mika, 133
mapping geographic data, 84
marketing
  analytics, 271–73, 293–94
  audience amplification, 289, 292–93
  branding, 290–91
  described, 288
  digestible content/multiple
    formats, 289–90
  email lists, 292, 293
  expert panels, 290–91
  omnichannel approach, 292–93
  podcasts, 290
  public speaking, 290
  scanners vs. readers, 289
  search engine optimization (SEO), 291–92
  social media, 289
  visuals, 289
Marshall, Adam, 151
Martin, Nina, 165–66
Martinez, Moises, 316
Mason, Betsy, 223, 225
mass incarceration, 179–80
mathematics
  abstract/concrete, balancing, 231–33
  derivations, 234–35
  described, 230–31

equations/jargon, 231–34
story selection, 233–34
visuals, 232
McCullom, Rod, 175, 178–79
mean/median, 33
media models
  career trajectories, 266–67
  missions in, 263–64
  newsroom trends in, 261–62
  niche, 262–63
  selection as career, 264–66
media science cafés, 321
medical journalism
  data collection, 148–49
  described, 147–48
  human element connection, 149–50
  interviews, 150
  legal issues in, 150–52
  verification, 152
mentoring programs, 320
Messonnier, Nancy, 237
metadata, 137–38
Metrics for News, 270–71
Metz, Cade, 197–98
Milankovitch cycles, 224
Miller, Greg, 173
misinformation, 281–82, 285–86
Mitjà, Oriol, 302
Mnookin, Seth, 192
mode, 33
Molteni, Megan, 195
Morbidity and Mortality Weekly Review, 157
Morisy, Michael, 118
Moskowitz, Clara, 206, 207
Most Unknown, The (Cheney), 70
Moynihan, Daniel P., 87
mRNA vaccines, 191
Mrs. Kelly's Monster (Franklin), 56
Muck Rack, 275
MuckRock, 123
multimedia
  applications of, 76
  data trails, 77
  history of, 74
  interviews, 76, 77–78
  nothing to see here problem, 78
  science journalism, 77–78
  story presentation, 77
  in story process, 76
  story rollout, 77
  target audience, 77
  training in, 75–77